An Islamic Court in Context

An Islamic Court in Context

An Ethnographic Study of Judicial Reasoning

Erin E. Stiles

palgrave
macmillan

First published in 2009 by
PALGRAVE MACMILLAN®
in the United States—a division of St. Martin's Press LLC,
175 Fifth Avenue, New York, NY 10010.

Where this book is distributed in the UK, Europe and the rest of the world,
this is by Palgrave Macmillan, a division of Macmillan Publishers Limited,
registered in England, company number 785998, of Houndmills,
Basingstoke, Hampshire RG21 6XS.

Palgrave Macmillan is the global academic imprint of the above companies
and has companies and representatives throughout the world.

Palgrave® and Macmillan® are registered trademarks in the United States,
the United Kingdom, Europe and other countries.

ISBN: 978–0–230–61740–7

Library of Congress Cataloging-in-Publication Data is available from the
Library of Congress.

A catalogue record of the book is available from the British Library.

Design by Newgen Imaging Systems (P) Ltd., Chennai, India.

First edition: November 2009

10 9 8 7 6 5 4 3 2 1

Printed in the United States of America.

This book is dedicated with great fondness to the late
Shaykh Hamid
and the staff of the Mkokotoni court

CONTENTS

ILLUSTRATIONS

Map

Photos

Tables

NOTE ON TRANSLATION

All interviews and court proceedings were conducted in Kiswahili, and I did all Kiswahili translations myself. Kiswahili terms are italicized. However, words that are used as titles for individuals (*Mzee, Bibi, Bwana, Shaykh, Mwalimu*) are not italicized because they are always used prior to the proper name. Although good manners require use of a title when addressing adults, I often refer to many people by name only to make a more readable narrative. I have, however, used titles for court workers, teachers, religious specialists, and for some elders. Most legal terms used in the text are derived from the Arabic but because they have become Kiswahili words, I use the Kiswahili spelling rather than the Arabic.

All names of individuals and villages have been changed for reasons of privacy. However, I have not changed the name of the court. I had full permission to conduct research there, and as there are very few primary courts in Zanzibar, it would be easily identified even if I changed the name.

I have included translations of many court documents, and have tried to stay as close as possible to the original. Court documents were handwritten entirely in Kiswahili, and then typed by clerks. However, if the *kadhi* cited the Qur'an, *hadith* literature, or an Arabic text he would handwrite the passage in the original Arabic on the typed document; I then copied the Arabic text by hand. I have noted when Arabic phrases were included in rulings by italicizing and I would like to thank Dr. Jocelyn Sharlet and Dr. Scott Reese for assistance with translating the Arabic portions of two rulings and in identifying the original source.

GLOSSARY OF
KISWAHILI TERMS

Bibi (Bi)	Ms.
Bwana	Mr.
Chuo (pl. *Vyuo*)	Qur'an school
Dawa	medicine, treatment
Dini	religion
Edda	a woman's mandatory waiting period after divorce before remarriage
Fasikhi	a divorce through dissolution by the court; no property changes hands
Haki	rights, legal rights, or duties
Hukumu	judgment/ruling
Kadhi	Islamic judge
Kadhi wa mtaa	local Islamic judge or religious expert
Kadhi wa wilaya	state-appointed Islamic judge
Khuluu	a type of divorce, often initiated by a woman, in which she financially compensates her husband
Kuandikia pesa	literally, "to write for money"; when a husband asks his wife for money to divorce her through repudiation
Kununua talaka	literally, "to buy a divorce"; used colloquially to refer to *khuluu*
Madai	plaintiff's claim document
Mahari	marriage gift from groom or groom's family to the bride
Majibu ya Madai	counterclaim of defendant, response to the *madai*
Masharti	terms or obligations in a ruling
Mchumba	fiancé, fiancée, potential marriage partner

Mdai	plaintiff
Mdaiwa	defendant
Mila	custom or normal practice
Mke	wife
Mkosa	the one who has made a mistake, he or she who is at fault
Mlezi	guardian
Mshahidi (pl. *washahidi*)	witness(es)
Mtaa	neighborhood, community
Mume	husband
Mwalimu	teacher
Mwari	girl who has reached puberty; young woman of marriageable age
Mzee (pl. *wazee*)	elder/elders, also a title for an older man
Mzungu	person of European descent; white person
Ndoa	marriage
Nia	intention
Rijista	refers to a registered divorce or the paper that indicates such
Shaykh	title of respect for a religious expert; used for *kadhi*s
Shamba	rural area, farmland
Sheha	government appointed community leader
Shehia	the political district/community over which a *sheha* presides
Sheria	law; religious law and Islamic law are *sheria za dini* and *sheria za kiislamu,* respectively
Talaka	divorce; also divorce by male repudiation
Ushahidi	testimony
Vyombo	household goods

ACKNOWLEDGMENTS

This project would not have been possible without the support and assistance of many people in Zanzibar and in the United States. Although all deserve credit for helping with the completion of this project, I alone am responsible for any mistakes or inconsistencies within. Fieldwork from 1999 to 2002 was supported by the Wenner-Gren Foundation and Fulbright-Hays. A follow-up fieldtrip in 2005 and writing in 2007–2008 were supported by Research and Creative Activity Awards from California State University, Sacramento.

Some material in this book has been published elsewhere, and it is used here thanks to generous permission of the publishers. A part of chapter three was published in *Africa. Journal of the International African Institute*, 75(4), Edinburgh University Press (2005). Sections of chapter two appeared in *Dispensing Justice in Islam*, edited by Muhammad Khalid Masud, Rudolph Peters, and David Powers (2005), Koninklijke Brill N.V. Much of chapter four was published in (2003) *Ethnology* 42(4). Map 1 was painstakingly prepared by Edward Allen.

To all my friends and neighbors in Zanzibar, thank you for putting up with me, answering questions, teaching me, feeding me, and making me laugh. My fieldwork was one of the happiest times of my life. I am indebted to all, especially to Mwamvua, Bi Salma and her family, Masoud, Abu Bakr, Mzee Mwadini, Bi Hadia, Moh'd, Kassimu, Mwalimu Ame, Bi Dawa, Mzee Fumu, Mzee Nyange, Mzee Faki, Bi Mwajuma, Mzee Haji, Babu wa Kikombetele, Bi Mwajuma, Ashura, Rukia, Asha, Mwanasharifu, and the rest of the girls. I think of all of you often. I would also like to thank the staff of the Zanzibar National Archives and the High Court for their assistance over the years. I am also grateful for the help of the *kadhi*s and staff with whom I worked at various courts.

I would like to thank the Department of Anthropology at Washington University, where this project began. Special thanks go to John Bowen for being both supportive and inspiring at every step. Thanks also to Ahmet Karamustafa, Jean Ensminger, Margaret Brown, and many other faculty members. I am particularly grateful to Priscilla Stone for helping make my research in Zanzibar possible when I could not go to Eritrea. Thanks also go to my fellow students, Brian, Vanessa, Beth, Sheela, Jen, Carolyn, Gareth, Laura, and Jason who made graduate school so rewarding. I am grateful to the Institute of African Studies at Columbia University for supporting me during a very valuable post-doctoral year. At CSU–Sacramento, I thank the members of the Department of Humanities and Religious Studies for taking a chance on an anthropologist, and I appreciate their collegiality and encouragement. Thanks also go to Afshin Marashi, Cindi Sturtz Sreetharan, Victoria Shinbrot, Wendy Matlock, and Kim Nalder for their friendship and support. I am pleased to have recently joined the Anthropology Department at the University of Nevada, Reno, and I look forward to many happy and productive years here with my new colleagues. Thanks also to Lawrence Rosen, who has been most helpful at various stages in my career, to Jenny Johnston for her enthusiasm for my projects and her visits to East Africa, and to Kathleen Kelly for advice and encouragement. Thanks also go to Lee Norton and Brigitte Shull at Palgrave for their support and assistance, and I appreciate the helpful comments of anonymous reviewers of this manuscript.

I'd like to thank my family for visiting me in Zanzibar and reassuring my friends there that I wasn't alone and adrift in the world. I'd especially like to thank my new husband Ed, who has been there from the frantic last minute photocopying of the dissertation to the last minute preparation of this manuscript. He has the patience of saint and is very good at making maps.

Finally, my most heartfelt thanks go to the late Shaykh Hamid, Bwana Fumu, and to rest of the court staff at Mkokotoni, to whom this book is fondly dedicated. I hope you will be pleased with this result of our efforts.

CHAPTER ONE

Kadhi, *Court, and Anthropologist*

One bright morning in December 1999, a young man named Abdulmalik came to an Islamic court in rural Zanzibar to ask for the return of his wife, Mariam. He told the judge, called a *kadhi* in Kiswahili, and the court clerks that Mariam left his home in a nearby village to live in Zanzibar Town with her sister. She refused to return to him because she claimed that he had divorced her through unilateral repudiation. A month earlier, he explained, Mariam had become angry with him one night when he was late returning home from the mosque after the evening prayer. He said that she was well educated in religious matters and was in the habit of giving him lessons in religion every night after prayer. They went to bed, and when he woke up in the morning, Mariam asked him if he knew how a man lawfully divorces his wife. He replied that he did not know, and she said him that she would teach him. She told him to get a pen and paper and to write the words "I, Abdulmalik, divorce Mariam, who will no longer be my wife, three times." Abdulmalik said that as soon as he had written the words, Mariam grabbed the paper, told him she was now divorced, and left. After hearing Abdulmalik's tale, the *kadhi*, a mild-mannered man in his sixties called Shaykh Hamid, recommended that he open a case, and Mariam was summoned to testify. When she came to court, Mariam confirmed the details of the case. She had indeed taught her husband to write the divorce paper, but she claimed that he had known that she wanted a divorce. In this case, it was up to the *kadhi* to determine whether the alleged divorce action was valid.

This book is about the way in which disputes such as this one are handled, arbitrated, and resolved in one rural court on the Zanzibari island of Unguja. It is also an intimate look at a working Islamic court and the everyday lives of the people in the surrounding community. I have conducted approximately 22 months of field research in Zanzibar between January 1999 and August 2008. Most of this time, I worked with Shaykh Hamid and his staff in a rural court and lived in a nearby village called Kinanasi. I also worked occasionally with other *kadhi*s in other courts in Zanzibar Town and elsewhere on Unguja. Specifically, this book examines the way in which judges, clerks, and litigants understand and use *sheria za dini* (religious law) in court cases involving marital disputes. In order to understand these disputes, I describe the relationship of local norms of marriage and divorce to what happens in court, and explore the role of the court and *kadhi* vis-à-vis the local community and the state, with a particular emphasis on the way in which the *kadhi* understands his position and his various obligations. Like his fellow *kadhi*s, Shaykh Hamid described his work as the final step in the dispute resolution process. He emphasized his role as a specialist who provided explanation of religious law to those who may not be well versed in it. Shaykh Hamid often told litigants, "The court is a hospital." He would explain that when a person was physically ill, he would go to the hospital for medicine to treat the illness. Likewise, when a couple experienced troubles at home, he would say, they needed to come to court to get the appropriate "medicine" to heal the marriage.

The Arabic term *shari'ah* is usually translated as "Islamic law," but does not correspond directly to law in the Western sense. It is rather a broader notion that addresses not only religious matters, but also provides divinely inspired regulation for many aspects of a believer's life, from the individual's relationship to God to social and commercial relationships between people. The Kiswahili term *sheria* comes from the Arabic, like many other Kiswahili words, but is used to refer to law in general; religious and other types of law are specifically designated: for example, *sheria za dini* means religious law, and *sheria za kanuni* refers to state law. Many countries with significant Muslim populations make provisions for Islamic law, Islamic courts, or both. Most often, Islamic courts have jurisdiction only over family law matters, and are frequently part of a state legal system that is not based on Islamic law.[1] In Zanzibar, Islamic courts are established as part of the state legal system and handle all family law disputes for Muslims. State-appointed *kadhi*s in Zanzibar are not bound by a family law code, as are their contemporaries in some other countries, but Islamic procedural law is circumscribed by secular state law.

Although the development of Islamic legal theory is a topic of interest to many contemporary scholars (e.g., Wael Hallaq 1997), the focus of this book is not on the foundational texts of the Islamic legal tradition. Nor do I spend much time discussing the particular texts that Zanzibari judges use to make their decisions, and I only reference them when the *kadhi* does. When I first worked with him, Shaykh Hamid most often cited the Qur'an or *hadith* literature (traditions of the Prophet Mohamed), and sometimes wrote decisions without reference to any sources at all (although somewhat later *kadhi*s were required to cite legal sources in all decisions). This was particularly in those cases he regarded as typical and thus formulaic. I am most interested in the social explanations of a *kadhi's* decisions, or how he uses, defines, and explains Islamic legal concepts with reference to and distinguished from other legal ideas. In my analysis, I take as a given the argument made by many scholars that legal reasoning cannot be understood without reference to wider cultural norms and the particular historical position of the courts and jurists. Shaykh Hamid himself reminded me of this on many occasions: he often said that given his position as an appointee of the Zanzibari state, he was neither able nor expected to apply *sheria za dini* in full. My analysis thus heeds Baudouin Dupret's recent call for scholars to move away from using Islamic legal doctrine as a "framing device" for studying legal practice, and to avoid considering legal systems as "samples" of wide-ranging legal models (2006). Dupret argues that scholars must avoid a top-down sort of approach because it is mistaken to assume that the way in which the law is practiced merely reflects how "the scholarly discourse on law is constructed" (146). Thus, researchers run the risk of telling court actors whether they are "properly" applying Islamic legal rules. Although Shaykh Hamid usually explained his decisions as in accord with Islamic law, from time to time he ruled with other ideas in mind, like fairness or penalty, that led him to work outside of his understanding of Islamic legal rules. A goal of this book is thus to demonstrate how people navigate between their own understandings of different sets of what John Bowen has called "legally-relevant norms" (1998a).

My analytical starting point is recent anthropological and historical work that suggests that Islamic legal traditions are open to interpretation and flexible and therefore must be studied contextually (Ewing 1988, Haeri 1989, Messick 1993, Bowen 1996, 2000, Tucker 1998, Peirce 2002, Peletz 2002, Zubaida 2006). These studies marked a significant move away from older views of Islamic law and legal traditions that considered it static and idealized, as in much Western scholarship on the subject (e.g., Joseph Schacht 1950). John Bowen writes that "most anthropologists and other Western scholars have taken Islamic law to be the rigid application of purportedly divine but strangely-superficial codes" (1996: 12) and proposes that anthropologists must instead understand jurisprudence as a cultural practice and interpretive process rather than a timeless body of rules: "*fiqh* [jurisprudential] practice is, and is supposed to be, socially and culturally variegated, taking into account custom as well as circumstance" (13). Furthermore, courts should be viewed as actively engaged in interpreting law, "The laws, the values and perceptions of the judges, and the powers of the court are all important in explaining the outcomes of decisions" (Bowen 2000: 124). Also, as we will see with Shaykh Hamid, judges sometimes go outside of religious law in permitting what they regard as technically unlawful practice in the interest of promoting fairness. To demonstrate, I consider patterns and modes of judicial reasoning, the *kadhi*'s perception of his role of judge between the community and the state, and the significance of gender norms and roles in legal activity.

Recent research on Islam and law in anthropology builds on the significant contributions of Lawrence Rosen, who like Clifford Geertz before him, is interested in culture as meaning. Rosen proposes that understanding judicial decision making requires acknowledging the local categories of meaning, and argues that judicial discretion should thus be regarded as a cultural phenomenon that is dependent on and a reflection of the cultural values in which the judge is situated. A decision is a product of cultural context because a judge's reasoning is embedded in cultural modes of thought; this is the "legal application of cultural assumptions" (Rosen 1989: 45). An anthropologist must consider the interrelationship of cultural assumptions, the legal approach, and substantive law. Rosen has addressed the process of judicial decision making rather than the law itself in cultural context, and others, like Bowen, expanded upon this by proposing the flexible and unfixed nature of law itself. Instead of looking at how cultural assumptions might inform judicial discretion, this suggests that law itself is transformable within the cultural context. This approach reflects developments in wider legal anthropology that, in opposition to earlier jurisprudential models, considered law not as a fixed entity but rather as processual and inseparable from the sociocultural context (Collier 1975, Moore 1978, Comaroff and Roberts 1981). As noted, the interaction of legal orders— religious or otherwise—has been of great interest to scholars, and in a move away from older ideas of legal pluralism, many have argued that different systems of law in plural societies cannot be viewed as independent isolates, but rather must be considered "mutually constitutive" (Merry 1988, 1991, Moore 1973). Such research suggests that coexisting legal orders, whether recognized by the state or not, interact and influence one another in discernible ways.

The coexistence of Islamic and other legal orders is thus a timely topic, when many states must address issues surrounding cultural and religious pluralism and the changing role of religious law and legal institutions. Although many scholars posit the flexibility of Islamic legal traditions, there have been only a few studies that look closely at the reasoning processes of actors in working courts, for example, the work of Rosen, Bowen, Mir-Hosseini (2000), and Peletz (2002). As Abdullahi an-Naim has noted, we know little of how "the shari'ah operates today" (2002: 19).[2] Dupret has also commented on the relative dearth of sociolegal studies from scholars in any discipline that attempt to situate the "law" in actual practices. He advocates a "praxiological" study of law, and suggests that instead of asking, "What is Islamic law?" scholars should attempt to answer the question, "What do people do when referring to Islamic law?" (2007: 81). He calls for researchers to consider "how people, in their many settings, orient themselves to something they call 'Islamic law' and how they refer personal status questions to the Islamic-law model" (82). The primary objective of this book is to do just this by considering the shape judicial reasoning takes with reference to Islamic legal ideas, the wider community, and the Zanzibari state. My focus is on the way in which the kadhi, clerks, and litigants conceive of and use Islamic legal principles and differentiate religious from other types of legal ideas and authority. Through relating the stories of several disputes, I demonstrate how court actors conceive of what is appropriately Islamic or un-Islamic, lawful or unlawful, and what is desirable, acceptable, or reprehensible in marital life and community behavior. As we shall see, the various actors reason based not only on their understandings of law but also their views of real and ideal marital behavior, local authority, and the role of the court in present-day Zanzibar.

As I have stated, this book does not seek to assess whether and when Shaykh Hamid conformed to Islamic legal principles as set out in standard legal texts, but rather how he and the other court actors understand and use religious law. I am, however, interested in the texts the court actors produce in the process of handling cases. In the anthropological study of Islam, a discourse-centered approach inspired by the work of Talal Asad (1986) has taken center stage in recent years. Scholars have attempted to bridge the gap between a focus in religious studies on normative texts and an earlier anthropological fixation on so-called little traditions by considering the production of particular discourses of religion, legal and otherwise, in particular sociohistorical contexts and with reference to normative texts and translocal religious practices (Eickelman 1985, Bowen 1993, Messick 1993, Mahmood 2005). In his work on the development of the use of legal texts and authority in Yemeni society, Brinkley Messick shows that Islamic "legal action does not occur in an isolated sphere but can only be understood in relation to the wider organization of socio-cultural life" (1990: 74). Messick asserts that *shari'ah* should not be understood as a fixed body of rules but rather as type of "total discourse":

> In treating the *shari'a* as the centerpiece of societal discourse, I place emphasis on the appropriation of its idioms, the flexibility and interpretability of its constructs, and the open structure of its texts.... I have moved away from an understanding framed in terms of the Western standard for law which has obscured the *shari'a*'s different range of social importance and its distinctive modes of interpretive dynamism. (1993: 3)

He describes the implications of literacy and the use of legal document as "social tools" and argues that "documentary practice" should be considered processually and in context. He demonstrates by considering the societal role of documents specialists because understanding their social identity is crucial to understanding the use of the legal documents.

As Dupret has pointed out, Messick's work does not focus on the "phenomenon of law practice and its work of producing formal records" (2007: 80), and my analysis attempts to connect the emphasis on written documents with the everyday practice of law. Another objective of the book is thus to illuminate the relationships between courtroom activity and the production of written documents. While working in court, I observed the way documents were produced by the court staff, and I copied or collected documents relevant to each case, most often the claim, counterclaim, and ruling (if any). Throughout the text, I draw on these documents. As historians Iris Agmon (2006) and Bogac Ergene (2004) have observed in their work on Ottoman courts, the way in which a case is framed and recorded in the official documents often differs markedly from how it developed in court. From their inception to their conclusion (if any), disputes are intentionally framed in particular ways at different stages in the court process by litigants, clerks, and the *kadhi*. Considering this, juxtaposing ethnography and documents is a valuable means of understanding how different parties prioritize particular issues and utilize Islamic principles and other values. The creation of documents shows what the writers consider important legal issues (for ethical or strategic reasons), how they exert control over the entextualization process, and how they understand the role of *kadhi* and court in community affairs.

As elsewhere in the Muslim world, women open the majority of cases in Zanzibar's Islamic courts, and a third objective of the book is to demonstrate how gendered experience of divorce and marriage influences court activity and the *kadhi's* reasoning. For many scholars, a common thread has been considering the ways in which women are often successful in asserting their claims and winning cases in courts that have been portrayed as primarily favorable to men (Antoun 1994, Hirsch 1998, Sonbol et al. 1998, Tucker 1998, Mir-Hosseini 2000, Peirce 2003, Agmon 2006, Stockreiter 2008). Some of these scholars have focused on the strategies female litigants take in presenting their claims to Islamic courts and striving for a beneficial outcome, and includes Peletz's work on Islamic courts and modernity in Malaysia, though gender is not his primary concern (2002). Richard Antoun found that Jordanian women generally draw on both their perceptions of Islamic law and local custom to maximize their position in court. For example, while family structure in the region is generally consanguine, the Islamic court tends to reinforce conjugal rights. As a result, women who have problems with consanguine ties have strategically turned to the court to support their claims against those bonds (1994: 57). Susan Hirsch researched *kadhis'* court disputes in the Swahili context of coastal Kenya, and argues that far from fulfilling the stereotyped image of passive victims of a patriarchal legal system, Swahili women find power and voice in courts that often decide in their favor. She observes, however, that women's use of the courts is not articulated as a quest for women's rights or a place where women can asset power over men, and complicates her own findings by noting that although women may achieve short-term benefit in the court from the *kadhi's* decision, there is a possibility of detrimental effects in the community by taking action. Considering this, one question that could use more discussion in these studies is what it means to "win" a case in court, and what the consequences of various decisions are even supposedly favorable ones. By moving back and forth between court and community, I hope to contribute to this body of scholarship by showing that it is difficult to simply describe court outcomes as "wins" or "losses" and that questions of gender, Islamic knowledge, and the application of Islamic law in courts must be examined with reference to how men and women in particular contexts experience and understand marriage, divorce, and family life.

Research in Zanzibar

I was somewhat surprised to find myself in Zanzibar in early 1999. I had planned to do fieldwork on court choice and forum shopping among Muslims in Eritrea but my plans changed when war with Ethiopia broke out in May of 1998. My advising committee and I settled on Zanzibar as a reasonable alternative: Islamic courts had jurisdiction over family law matters for Zanzibar's majority Muslim population and we thought that because of Zanzibar's historical connections to Oman and Yemen, my years of studying Arabic might be useful there. Once the decision was made, I started studying Kiswahili and frantically began to read about Zanzibar and the Swahili coast. When the granting agencies that had agreed to fund the Eritrea project approved the change of field-site and refocusing of the project, I left St. Louis on a clear, icy morning for hot and humid Zanzibar.

Today, Zanzibar is a semiautonomous island state of Tanzania with its own president, parliament, and semi-independent legal system. The Zanzibar archipelago consists of two major islands, Unguja and Pemba, and several smaller ones. The islands are often described as part of the wider Swahili Coast of East Africa, extending from Mozambique to Somalia. Indeed, Kiswahili is spoken everywhere on the two islands; it is the first language of most Zanzibaris, although there are significant variations in dialect from one region to the next that many people delighted in explaining and demonstrating to me. Zanzibar's population of about one million is over 95 percent Muslim, and most people follow the Shafi'i *madhhab*, or legal school.[3] About 70 percent of the population lives on the southernmost island of Unguja. There is one fairly large city on Unguja, Zanzibar Town, with a population of perhaps 200,000. There are no other large urban centers on Unguja, and outside of Zanzibar Town most people live in very small towns and villages of various sizes.

I conducted most of my research in rural northern Unguja. However, when I first arrived in early 1999, I spent about three months living and working in Zanzibar Town. Thanks to the help of an American teacher who had lived in Zanzibar for many years, the late Dr. Dennis Doughty, five days after my arrival I had a private Kiswahili tutor and had moved in with a large family. My new mother, Bibi Serena,[4] was a plump and exuberant woman of about 60, who presided over a lively household of unmarried and divorced daughters and several grandchildren. Over the years, the family had hosted several students of Kiswahili on in their breezy two-bedroom apartment in the Michenzani neighborhood to the east of Zanzibar's historical Stone Town. I eased comfortably into life

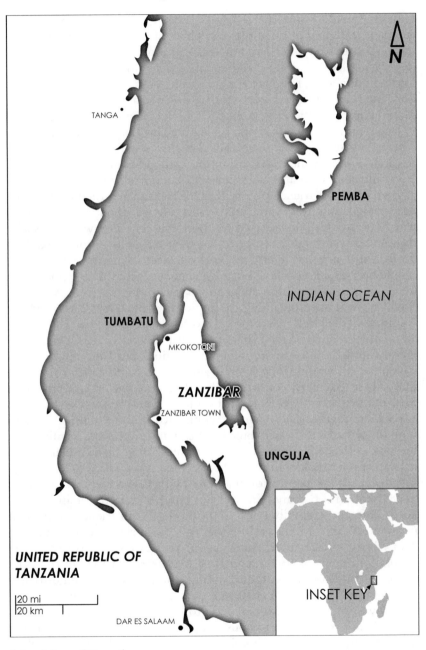

1.1   Map of Zanzibar

with Bi Serena's family for a couple of months while I improved my language skills and planned the next phase of the project. My time was taken up with studying Kiswahili, learning more about recent Zanzibari history and legal traditions, and choosing a site for my future research. During this period, I also collected appeals records from the High Court files, interviewed the Chief *Kadhi* and other legal specialists, and made several trips with the Zanzibar Legal Services Center to rural areas for village seminars on human rights and state law. Like most other foreign scholars, I based myself at the National Archives, which had arranged my research permit while I was still in the United States. On my first day, I was introduced to a young man who became an invaluable research adviser and a fast friend. Bwana Rashidi had just returned to Zanzibar from completing his BA in Saudi Arabia. He was about my age, which was 27 at the time, and seemed to know everyone in town—especially those who were highly respected for their religious learning. Rashidi decided that my first interview on the subject of marriage and family should be the Chief *Kadhi* of Zanzibar, and he scheduled an appointment for us to meet him within about three weeks of my arrival. After that initial meeting, and obtaining the approval of the Chief *Kadhi* and the Registrar of the High Court for my project, Rashidi and I began visiting all six *kadhi's* courts on the island of Unguja. When I arrived, I knew that there were Islamic courts throughout Zanzibar, but I did not know in which court I would work. The largest court is in Zanzibar Town, and is usually staffed by two or three *kadhis*. There are five others in the rural parts of the island, all easily reachable on the converted trucks that serve as public transport, *gari ya shamba*, within two or three hours. Although most of the *kadhis* expressed some interest in my project and a willingness to work with me, none matched the enthusiasm of Shaykh Hamid and his staff at the Mkokotoni court. My decision was made.

After receiving permission from the High Court Registrar to work in Mkokotoni, I prepared for the next phase of the project and moved north. After several jolly house-hunting trips with friends from town in a rented van, I found a white-washed coral-brick house in the verdant, shady village of Kinanasi, about two miles by road from the Mkokotoni court. I had intended to live nearer to the court in the village of Mkokotoni itself, but finding a house there proved difficult. Furthermore, I faced the challenge of a friendly and helpful regional officer who strongly encouraged me to live in one of the three "bank houses" built years earlier to house employees for a new bank that never opened. Unlike most other homes in the area, the bank houses had electricity, running water, and furniture; at the time, two were inhabited by two young women, an American Peace Corps volunteer and an English VSO volunteer, and the officer thought that I should rent the third. Despite the attractions of water and electricity, the houses were in a somewhat isolated fenced compound, and I imagined it might be difficult to establish relationships in the surrounding villages if I lived there. I finally convinced the officer that I would be just fine living in a "local" house, as he called it. As a compromise, I agreed to live in the bank house for one month while we made minor repairs on the coral-brick house. At my request, his recently divorced niece Bi Mwanahawa had agreed to live with me, and as she was a smart and accomplished woman from the community, I think that helped convince him that I would survive in "local" accommodations.

After I moved north, my work consisted of days spent in the Mkokotoni court and out-of-court research on marriage, divorce, and norms of daily life. I attended court every Monday through Thursday for over 13 consecutive months in 1999 and 2000; in 2002 and 2005, I also spent several more weeks in court. Court research included listening to all open cases as they came up, discussing cases and legal problems with the *kadhi* and the clerks, conversations and interviews with litigants, and perusing court files on *kadhi*'s cases from the past 20 years. In many cases, I conducted follow-up interviews with litigants outside of the court. Shaykh Hamid and the clerks welcomed me warmly to the court, and from my first day I had my own chair between the clerks' and the *kadhi*'s desks. The High Court, one of the many offices from which I needed research clearance, had specifically declined to give me permission to tape-record actual court sessions, and thus I took very detailed notes on a laptop instead. Although at first I was concerned about not recording court sessions, I eventually realized it was for the best. My Kiswahili skills improved rapidly, and since I had no back up tape to rely on, I had to concentrate with great intensity during every court session; when I realized the *kadhi* and clerks welcomed questions, I was not shy about asking for clarification when I was confused. I was, however, able to record my interviews with the *kadhi* and clerks and with some of the claimants. In January 2000, after working in Mkokotoni for about seven months straight, I began to spend one day per week in the primary court in the village of Mfenesini with the dynamic *kadhi* Shaykh Vuai Omari. Mfenesini was about halfway between Mkokotoni and Zanzibar Town, or about 30 minutes by the *gari*. I also worked a bit with Shaykh Vuai in research trips in 2002 and 2005, and although I do not explore any of his cases in detail here, I do occasionally compare his reasoning processes to those of Shaykh Hamid.

Decades ago, Laura Nader and Barbara Yngvesson argued that to understand legal practice in the courts, anthropological studies of law must not focus solely on court practice but must also consider the litigants themselves and the historical role of the courts in the wider community (1973). To this end, I did extensive research in Kinanasi and the wider community surrounding the court concerning locally held ideas about marriage and divorce, dispute resolution, religious practice, and how people gain knowledge about the law. This information was collected through innumerable informal discussions with neighbors and friends as well as semistructured interviews with many divorced women and men. I also interviewed religious teachers, local religious experts, and *sheha*s, who are government appointed community leaders. In addition to interviews, participant-observation at weddings, funerals, birth ceremonies, religious celebrations, women's Qur'an classes, and other activities contributed to my understanding of social and cultural life.

The Courts in Recent History

Although this book does not seek to provide a comprehensive history of *kadhi*s courts or Islamic law in Zanzibar, it is useful to review some developments in the recent history of the courts.[5] Reforms enacted in the colonial period were particularly influential, and have had a significant impact on the shape and function of the *kadhi*s' courts today. Furthermore, a brief review of the position of *kadhi*s in recent history shows that the position of *kadhi*s between different, and sometimes competing, elements of society is not unique to the present day.

*Kadhi*s have arbitrated disputes in Zanzibar since at least the early part of the nineteenth century, and probably long before, though few sources are available before the nineteenth century. Throughout much of the nineteenth century, Zanzibar was a part of the Omani Sultanate along with the coastal regions of present-day Tanzania and Kenya. The Busaidi Sultan of Muscat, Seyyid Said bin Sultan (1804–1856), made Zanzibar his capital in 1832 after suffering resistance and defeats at Mombasa against the Mazrui Arabs (Lofchie 1965, Middleton and Campbell 1965). Zanzibar then became an independent Arab state, and remained so until Britain made it a protectorate in the 1890s. During this period, *kadhi*s were available to adjudicate matters at the local level, and the Sultan acted as the final court of appeal. Randall Pouwels has proposed that before the reign of Said, it was likely local governors, *liwalis*, rather than Islamic legal experts who were administering the law in Zanzibar (1987). Although John Gray writes that Seyyid Said "tried his hand" at judicial reform and went out of his way to try to appoint *kadhis* of worthwhile reputation who were respected in the community (1962), little is known about Seyyid Said's specific legislation affecting the islands, and no formal process of enacting legislation is evident. The one significant piece that is considered to be a decree from the period of Said, dating back to 1845, stated that each *kadhi* must resolve a dispute according to his own school of Islamic law. However, Gray suggests that since the decree referred to this practice as a remainder from "the old times," it does not represent a legal innovation but rather a declaration of the existing law (1962).

At this point, most of Zanzibar's population followed either the Ibadhi or the Sunni Shafi'i *madhhab*. The Busaidi Sultans and the Arab population, made up of immigrants from Oman and their descendants, generally were Ibadhi while the rest of the population was generally Shafi'i; this included people of mainland African descent, among them many slaves, and those who identified as Watumbatu, Wahadimu, Wapemba, or Shirazi. Seyyid Said appointed one principal *kadhi* each for the Shafi'is and the Ibadhis, and they and other *kadhis* worked somewhat informally as there were no specified court buildings or offices. *Kadhis* likely heard disputes at their homes and Gray suggests that as late as 1891, some *kadhis* in Pemba were holding court in the streets of Chake Chake, one of the larger towns (1962). Under Said, the policy of noninterference in local custom took precedence to the extent that *kadhis* could be paid and appointed by the Sultan, but they had to secure the local elders' support (Pouwels 1987).

After Seyyid Said's death in 1856, there was a dual succession in which one of his sons ruled in Oman and another in the islands and coastal region of the African mainland. Eventually, the kingdom was politically divided. Seyyid Majid, Said's second eldest son, ruled Zanzibar from 1856 until 1870. Majid's brother, Seyyid Barghash, succeeded him and became Sultan of Zanzibar from 1870 until 1888. In a series of developments that corresponded with increasing British interest in the region, Zanzibar eventually became fully autonomous from Muscat. For example, Seyyid Thuwani, who succeeded Said in Oman, was unhappy with Majid's rise in Zanzibar and prepared to take it militarily, but was thwarted by British intervention (Gray 1967). In 1884, during European expansion in Africa, Britain and Germany signed a treaty that deprived the then Sultan of Oman of most of his possessions in East Africa, save for a coastal strip that included Zanzibar.

While Said and his immediate successor Majid were reputed to have maintained a certain distance and detachment from their non-Arab subjects in Zanzibar, sources suggest that Barghash made active inroads into influencing non-Arab communities, and made more significant attempts than his predecessors to intervene in local level religious and judicial matters. In at least part of the Sultanate period, it appears that the relationship between Shafi'i and Ibadhi leadership was respectful; B.G. Martin writes that Shafi'i and Ibadhis appeared to be on the "same footing" in the *barazas*, or regular meetings, of the Zanzibar Sultans (1971). He notes, however, that this was more prominent under Said than Barghash, and more of the *ulama* under Said and Majid were Shafi'i. Barghash divided his territory into small polities, *wilayat*, over which *liwalis* presided with some power to resolve disputes. He also sent Ibadhi *kadhis* into the non-Arab interior of the islands of Zanzibar, which were populated primarily by adherents of the Shafi'i *madhhab*. Some scholars suggest that a form of customary law was prominent in these regions at the time, and John Middleton and Jane Campbell write that, consequently, "the original diversity of customary law was greatly modified with the rigid application of the *shari'a* since Sultan Seyyid Barghash sent *kadhis* into the Shirazi areas" (1965: 19). Pouwels notes that Barghash implemented a program through which he had much more control over the *kadhis* in outlying areas, "he established a more formal judiciary system in Zanzibar which defined more clearly demarcated areas of jurisdiction under his many court *qadis* and *ulama* and put them all on the state payroll" (1981: 333). Martin, however, states that while we do know they were on the payroll, it is difficult to determine what this meant, exactly; we do not know, for example, what they were paid or whether there were differentials in pay for different types or levels of *kadhis* (1971).

Under Barghash, then, Zanzibar saw an increase of Ibadhi authority that extended into rural areas of Unguja and Pemba, as he appointed Ibadhi *kadhis* but also faced the conversions of Ibadhi scholars to Sunni Islam and the Shafi'i *madhhab*. Pouwels writes that the conversions were based on the precariousness of the early Sultanate and the "loose means which it employed to retain the allegiance of the many coastal locations" (1987: 116). Also, it seems that many nominal Ibadhis had only a little education and no real connection to Oman, and most stopped using Arabic in day-to-day life within a couple of generations of moving to Zanzibar. Essentially, there were few discernible differences between the Ibadhis and the Shafi'is, and "it is for such reasons, therefore, that many Ibadhis, even the families of the liwalis, like the Busaidi of Lamu, began converting to Shafi'i Islam after a generation or two" (Pouwels 1987: 116). Pouwels also comments on the degree of social interaction between Omanis and Swahilis in towns, and notes that by 1907, most people in Zanzibar were Shafi'i. An important source on *kadhis* in the Sultanate and protectorate periods comes from the writings of Shaykh Abdallah Salih Farsy, who is well known for his Kiswahili translation of the Qur'an and his writing on legal matters (Elmasri 1987). In his hagiographic *The Shafi'i Ulama of East Africa, 1830–1970,* he writes about the occurrence of conversions from Ibadhi to Shafi'i by some *kadhis* and scholars. One Shaykh Barwani so infuriated Barghash when he became a Sunni Muslim after conversations with Shafi'i scholars that Barghash imprisoned him and then exiled him to Oman. It is evident that local forms of authority changed throughout the period: *kadhis* became salaried appointments of the Sultan, and village elders were subsidized as well, which Pouwels describes as "buying the loyalty" of subjects (1987). Whereas in the past, town elders had much say over who had legitimate authority to resolve disputes, local level leaders in this period were appointed by the Sultanate, and were thus Ibadhis, not Shafi'is. However, Allyson Purpura writes that even though *kadhis* were appointed as Ibadhi experts by the Sultan, other kinds of religious expertise and Islamic knowledge were still important, for example, the Sufi shaykhs in Zanzibar Town (1997). Gray addresses perceptions of *kadhis* in the nineteenth century, and refers specifically to British Consul Hamerton, who wrote about the lack of impartiality and general disrespect the population had for the *kadhis* in the mid-nineteenth century, an opinion with which his successor Rigby concurred (1962: 145). However, later historians have criticized this view, noting instead that the allegations of corruption likely stemmed from the fact that British administrators were not knowledgeable about local norms of exchange and gift giving. Pouwels, for example, writes that foreign officials often interpreted the practice of giving gifts of thanks to the *kadhi* as instances of bribery (1987).

Much like today, community leaders known as *shehas* also held authority in Zanzibari communities in the pre-Sultanate and Sultanate periods. Political districts were divided into *shehias*, which in the rural areas consisted of groups of several villages, each of which was presided over by a *sheha*. In the northern region of Unguja, where I conduct research, the Watumbatu people were governed by a *sheha* who was under some authority of a ruler known as *mwinyi mkuu* until the mid-nineteenth century. The *mwinyi mkuu* was the historical leader of the Wahadimu people of central Unguja, who presided over a hierarchical system of local government and had legal authority (Ingrams 1931). Historians have suggested that female rulers were prominent in the northern part of Unguja, and the woman know as the last Mtumbatu queen, Mwana wa Mwana, was married to a *mwinyi mkuu* called Hassan (Gray 1977, Nuotio 2006). The historical record suggests that in the nineteenth century it was likely that the *mwinyi mkuu* appointed *sheha*s, although there was some attempt to maintain the appearance that the communities themselves selected shehas. The *mwinyi mkuu* asked people whether "the candidate was the man they wanted as *sheha*, and the people would dutifully reply in the affirmative" (Gray 1963: 165). If he did not make the outright appointment, then the candidates were likely subject to his confirmation and approval (1963). Although the last *mwinyi mkuu* died in 1865 and no other was appointed (Ingrams 1931), *sheha*s have remained influential throughout Unguja and Pemba; Elke Stockreiter notes their role as mediators in *kadhi*'s courts during the colonial period (2008), and I will take up their role today in later chapters.

In 1890, Great Britain made an agreement with Seyyid Barghash that legally created the protectorate in Zanzibar. A dual mandate was established with, essentially, two heads of state: the Sultan and the Queen of England. Britain was to control only foreign relations, while the Sultan maintained control over domestic matters. Stockreiter argues that despite the colonial goal of preserving the "idea" of an Arab state through upholding Islamic law as fundamental and through avoiding interference in substantive Islamic family law, reforms in the period significantly impacted the jurisdiction of religious law (2008: 44–45). Many of the reforms of the legal system in the protectorate period remained intact until after independence in the early 1960s, and many are also influential in the present-day legal system.

In 1897, the first district courts were established and a dual system of law that made provisions for British and Sultanic law was set in place (Vaughan 1935, Allott 1976). That same year, the Sultan Hamud bin Muhammad issued a decree that divided Zanzibar into three districts: Mkokotoni, Mwera, and Chwaka, each of which was headed by a *liwali* and had a *kadhi* to hear civil disputes (Stockreiter 2008). A decree of 1899 established rules for procedure in the courts, which required the hearing and recording of the testimony of witnesses, and required *kadhi*s to give reasons for decisions; a 1904 regulation required cases to be heard in the district in which they arose. A 1908 decree gave British judges more authority in the Sultan's courts and gradually moved toward an assimilation of the two jurisdictions (Vaughan 1935, Stockreiter 2008). In 1923, the British Subordinate Courts Order and Zanzibar Courts Decree were passed. This legislation established five classes of subordinate courts, including *kadhis* courts, and moved toward merging the two jurisdictions. All of the courts were to have criminal jurisdiction save those of the *kadhi*, which heard all other matters in which both parties were Muslim. All of the subordinate courts had appeal to the British courts (Anderson 1970). The 1923 decree declared Islamic law to be the fundamental law of the protectorate, although this was limited by Sultanic decrees concerning criminal law and other regulations concerning procedure, like the 1917 Evidence Decree (Stockreiter 2008). In the British courts, common law ruled. Ibadhi and Shafi'i law both were to be administered in the Sultan's court, and Anderson writes that litigants were able to choose between *madhhabs* (1970). However, as Stockreiter notes in citing Vaughan (1935), this was unlikely because only one district in 1935 had both *kadhis* (2008). Also, British appellate judges were required to use the school of the litigant save in property cases—then, the school of the *kadhi* who originally decided the case must be followed in the appeals. If the *madhhabs* of the vying litigants were different, then the judge was to follow whichever *madhhab* seemed more equitable in that situation.

Significant amendments to the 1923 courts decree were introduced in 1947. The jurisdiction of the Islamic courts was officially limited to personal status, including marriage, divorce, and inheritance cases in which less than 1,500 shillings was at stake. Civil proceedings in which the estimated value of the property at hand was not over 800 shillings were also subject to the Islamic courts. Anderson notes that, by this period, customary and religious law had come to be linked in reference to family law, but that differences remained in laws concerning land tenure (1970). Much legislation from this period remains influential today as received law. The 1917 Evidence Decree required witnesses be heard without being questioned about their religion, and this is maintained today as the basis for the evidentiary rules spelled out in the *Kadhi's* Courts Act of 1985, which prohibits discrimination against witnesses and their testimony based on religion, gender, or ethnicity. The same decree established documentary evidence as legitimate and specified that witness testimony of those who have interest in the case should be rejected.[6] In a parallel to today's courts, Anderson writes that many *kadhis* refused to recognize decrees regarding rules of evidence (1970: 69), and Stockreiter describes a "clash of values" concerning procedural law on witnessing (2008: 81).

The Marriage and Divorce (Mohammadan) Registration Decree of 1935 and 1936, amended in 1944, provided that all marriages and divorces and revocations of divorce be registered with local authorities. In the original decree, the courts were not to recognize unregistered marriages, but this was repealed in 1944. Also, it stated that a woman could sue for back maintenance in the Shafi'i school but not in the Ibadhi. In this period, neither Ibadhi nor Shafi'i judges could grant a divorce for reasons of a husband deserting his wife unless it could be proved that he did not leave adequate maintenance. Also, a judge might simply order a man's property to be seized and sold in order to support his wife if it was deemed he was not maintaining her adequately. Accusations of adultery were designated as penal matters, although no Islamic criminal law remained (Anderson 1970); adultery is a matter for criminal courts today.

Zanzibar was granted internal self-government in 1963, and the first constitution came into effect with independence in December of the same year. At independence, the Sultan remained head of state, but a bloody revolution overthrowing the Sultanate government followed in early 1964. In the same year, Zanzibar joined with newly independent Tanganyika to form the United Republic of Tanzania, although Zanzibar has remained semiautonomous. A revolutionary government was established in Zanzibar under the leadership of first president Abeid Karume. Throughout the British period and with the terms of independence, Zanzibari Arabs maintained their domination of political life and local government. Some scholars propose that the continued dominance and the favored status of Zanzibari Arabs by the British was the major cause of the Zanzibar revolution (Lofchie 1965). However, others argue that this ethnic conflict model is a simplistic explanation for a more complex phenomenon; Abdul Sheriff, for one, argues that ideologies of race and ethnicity "were allowed to take precedence," and thus masked the more substantial underlying class conflicts (1991). Decree No. 1 of 1964 had maintained the British and Indian laws and the court system that had been established during the protectorate. Islamic courts had wide jurisdiction in this period and heard any civil suit in which both parties were Muslim. However, in 1966, the British and mudirial courts were abolished, and a new Courts Decree established a High Court, and magistrate, *kadhi*, and juvenile courts (Bierwagen and Peters 1989).

One of the most interesting developments in the postindependence period was the complete, but short-lived, overhaul of the legal system introduced by the radical People's Courts Decree of 1969, introduced during Zanzibar's experiment with socialism. In a dramatic break from the preceding legal system, the decree established nine People's Area

Courts on Pemba and Unguja, People's District Courts, a High Court, and a Supreme Council. Little has been written about these courts, and I found people reticent to talk about them; the courts do not seem to be remembered fondly. It is clear, however, that all of the primary courts were presided over by party chairmen appointed by Karume; they had no particular legal training (Ramadhani 1978–1981) and Hank Chase notes in an article on the treason trial following the 1972 assassination of Karume the likelihood that many of these individuals were only semiliterate (1976: 23). Furthermore, there were no rules of evidence or procedure, and advocacy was not allowed (O. Sharif, n.d.); Chase notes that the prosecutor served as the defendant's counsel (1976).

Although Bierwagen and Peter, citing A.S.L. Ramadhani, write that the People's Courts Decree also established *kadhi*'s courts, many Zanzibaris told me that there were no provisions for state-supported Islamic courts during this time. And although community religious experts continued to exercise authority over religious matters, many people explained that during this period people were required to go to the local ruling party "chairmen" with marital problems or to get a divorce. Despite this, no one told me of a time when religious law was not at the very least nominally adhered to in marriage and divorce. Although the People's Courts are certainly worthy of note in a discussion of Zanzibar's recent legal history, they have little relevance to today's legal system since they were entirely eradicated in the 1980s.[7] The existing system was instituted in 1985, in what Bierwagen and Peter have termed a "silent revolution" in the organization of the judicial system (1989). This is an apt description since the People's Courts were eliminated and replaced by a return to common law in a variation on the protectorate system.

The 1984 constitution established a new High Court of Zanzibar and two types of subordinate courts: magistrates' and *kadhi*'s courts. The *Kadhi*'s Courts Act of 1985 established Islamic courts at both the primary and appellate levels. The Chief *Kadhi* was instituted as appellate judge, and final appeal rests with the Chief Justice and a council of four religious scholars. Unguja is divided into six administrative regions, each of which has an Islamic primary court. Pemba, is divided into three. There are nine primary Islamic courts in Zanzibar today; three on Pemba and six on Unguja, and litigants are expected to go to the court that has jurisdiction over their political region. There is, however, some leeway for those who request to be heard in different districts for the sake of convenience. Table 1.1 shows the distribution and regional jurisdiction of the Islamic courts.

Table 1.1 Zanzibar's Islamic Courts

Court, Location	District Served
Chief *Kadhi*'s Court; High Court Zanzibar Town, Unguja	All appeals—Unguja and Pemba
Unguja's Primary Kadhi *Courts*	
1. Makunduchi	South
2. Kariakoo	Zanzibar Town
3. Mwera	Central
4. Chwaka	Central East
5. Mfenesini	North B
6. Mkokotoni	North A
Pemba's Primary Kadhi *Courts*	
1. Konde	Central
2. Wete	North
3. Mkoani	South

Today, the codified laws relevant in Zanzibar include British and Indian laws, which are considered applicable until replaced by local law. In addition to imported law, the constitution establishes that different types of local laws are also authoritative. These include those Sultanic decrees from the past that were countersigned by the British Resident, decrees of the revolutionary council, customary law, union legislation, and Islamic law. The constitution specifies that protectorate law and Sultanic decrees make up the body of received law unless they are overruled or limited by later legislation. Since Zanzibar is semi-autonomous, Tanzanian union legislation is only applicable to those matters deemed of interest to the union. Because of this, the Tanzanian Marriage Act of 1971, which established a minimum age at first marriage among other regulations, is not applicable to Zanzibar (see Hashim 2006). Accordingly, decisions of the Islamic courts in Zanzibar are not considered relevant to the Tanzanian union and thus are not appealed to the Supreme Court of Tanzania. Rather, final appeal rests with the Chief Justice of Zanzibar. *Kadhi's* courts have jurisdiction over family and personal status matters. Although substantive family law remains relatively untouched and there is no family law code, procedural law is moderated by the Evidence Decree of 1917, which required witnesses be heard without being questioned about their religion. This is maintained today as the basis for the evidentiary rules spelled out in the *Kadhi's* Courts Act that prohibits discrimination against witnesses and their testimony based on religion, gender, or ethnicity. Also in

1985, an act known as the Spinster and Divorcee Protection Act was passed, which subject an unwed woman to two years in prison if she became pregnant; the father of the child would also be imprisoned for up to five years (Fluet et al. 2006).[8] The law was repealed in 2005 due to serious criticism and replaced with a new act. In a 2005 interview from the Zanzibar Female Lawyer Association (ZAFELA), the lawyers agreed that the new law serves to protect unmarried women and children born out of wedlock with secular laws, since they were not protected adequately through *sheria za dini*. The replacement statute does not require a prison sentence but requires unwed mothers to perform community service. Unlike the previous act, the new law also permits a woman to return to school after delivering the baby. Fathers are required to support the child, and also perform community service (Fluet et al. 2006).[9] Issues relevant to the custody of these children would be handled in the secular rather than the *kadhi's* courts.

During the time of my research, all of the *kadhi*s in Zanzibar's primary courts were nominally appointed by the president of Zanzibar, though they were recommended by the Chief *Kadhi* and then underwent an interview and examination process; although the *Kadhi's* Act cites the authority of a Judicial Service Commission to appoint *kadhi*s in conjunction with the Chief *Kadhi*, I did not hear reference to this body (Bierwagen and Peter 1989: 407). Although *kadhi*s are of course required to be Muslim, there was neither standard training for *kadhi*s nor specific educational prerequisites for achieving the post. Rather, they were appointed based on their reputation as religious scholars and as a result had fairly diverse educational backgrounds; many were educated abroad in places like Saudi Arabia and Egypt, although the *kadhi* with whom I worked most closely studied solely in Zanzibar and Tanzania. Although all of the state-appointed *kadhi*s follow the Shafi'i *madhhab*, most claimed to make provisions for disputants from other *madhhab*s when and if they appeared in court. *Mwalimu* (teacher) Simai, a local religious teacher who also acts as an informal *kadhi*, once told me emphatically that he and the state-appointed *kadhi*s would absolutely attempt to resolve matters through another *madhhab* if asked, and in his view, effective legal scholarship relied on frequent reference to other schools. However, avowed followers of other schools are few in Zanzibar, and are especially rare in rural areas. In my time in court, I did not hear anyone ask for a *kadhi* to consider another *madhhab*.

Since my initial research period in 1999–2000, there have been changes relevant to the *kadhi's* courts. By 2002, *kadhi*s were required to write their rulings in a more formulaic fashion akin to rulings in the civil courts, just as the plaintiff's claim and defendant's counterclaims already mirrored their secular counterparts. Furthermore, *kadhis* were asked to cite the legal source for their ruling in every decision. In a 2002 interview, Shaykh Hamid told me that he was working every Thursday with a new *kadhi*—the one who eventually replaced him at Mkokotoni after his death. When I asked him why they worked together, he gave the following explanation:

> Before giving the ruling to the plaintiff and the defendant, we work together to scrutinize the law, to make sure that we did it right [came to the appropriate ruling]. It is to make sure that we have decided the cases to make sure everything is fair and right... On Monday, Tuesday and Wednesday we don't give *hukumu* (rulings) at all—not until Thursday [until after they have met together].

He continued by explaining that they wrote all of their decisions together, and waited to read them aloud to litigants until they had agreed on the ruling.

In 2001, a bill was passed making official the *fatwas* of the *mufti* of Zanzibar, and this has the potential to influence judicial activity at the primary level. Although a *mufti* had operated in an unofficial capacity for several years, the bill legalized the office of the Chief *Mufti*.[10] Among others duties, the bill gives the Chief *Mufti* the duty to mediate disputes between the government and Muslim organizations, stipulates that his legal opinions are binding, and calls for Muslims to seek the *mufti's* opinion before holding communal prayers. As of 2005, the law had no discernible effect on activity in the primary courts, and in my interviews from that year, some *kadhis* scoffed at the idea that the *mufti* could have any authority over their decisions or court practice. However, if the new emphasis on standardized written *hukumu* is any indication, it is possible that there will be increasing oversight in the primary level courts.

Courts and Community Today

Although the courts handle a range of problems concerning marriage and divorce, child custody, and inheritance, the vast majority of cases opened concern marital disputes.[11] Divorce is very common in Zanzibar today, and apparently was also not unusual in the recent past. In the mid-twentieth century, J.N.D. Anderson noted that a Shafi'i *kadhi* told him only about 20 percent of Zanzibari marriages did not end in divorce (1970), and Stockreiter confirms the frequency of divorce in the early twentieth century (2008).[12] Unlike many states (e.g., Egypt, Indonesia, Eritrea, Tunisia) in which state-level legal reforms have limited the right of men to divorce their wives through repudiation (Esposito 1982, Watson 1992, Moors 1999), men in Zanzibar maintain the unmitigated right to unilaterally divorce their wives through repudiation without the consent of the wife or the court. Women may file for divorce in court and are granted it on a variety of grounds, the most common being desertion and lack of maintenance. However, most divorces happen out of the court through repudiation or variations on a type of divorce known as *khuluu* (also written as *khului* or *khula*, from the Arabic *khul'*), in which a woman compensates her husband financially; this is common in Zanzibar and will be discussed at length in later chapters.

The Mkokotoni court serves the northern district of Unguja, which is a rural region of nearly 100,000 people (photo 1.1). The village of Mkokotoni is situated in the northwestern part of the island, and is the administrative center for northern Unguja. The rural areas are referred to collectively as *shamba*, which means farm or farmland. Like the rest of Zanzibar, this is not a wealthy area, and most people are quite poor. Zanzibar's 2004/2005 Household Budget Survey cites a poverty rate at about 50 percent, and the gross domestic product per capita at about 300 USD, which was slightly higher than 2002 figures. Most people in the region subsist through small scale farming, augmented with fishing, running small businesses like shops, or participating in the informal economy. In this part of Unguja, the majority of people identify ethnically as Tumbatu (sing. *Mtumbatu*, pl. *Watumbatu*). All litigants are asked to state their *kabila* (ethnicity) when preparing court documents, and I only rarely heard people describe themselves otherwise in court or elsewhere. Scholars have often described the Watumbatu as one of the original peoples of Zanzibar, along with the Wahadimu and Wapemba (Campbell 1962, Trimingham 1964, and Middleton 1992). Many Zanzibaris who identify as such, like the Tumbatu people I worked with, describe their ancient origins in Shiraz, sometimes via Somalia or Ethiopia. Many people in Zanzibar also identify primarily as Shirazi to this day. Beyond simply asking litigants to identify their ethnicity for preparing claim documents, mention of ethnicity almost never entered discussions in the courtroom or my conversations on marriage and divorce out of court, and thus there is little further reference to it in this work. Although almost all of my acquaintances identified as

Photo 1.1 The Kadhi's Court at Mkokotoni lies through the door under the archway. The regional administration building in Mkokotoni houses the Kadhi's Court.

Tumbatu when asked, no one ever discussed "being Tumbatu" as significantly different from any other ethnicity in Zanzibar. References to linguistic idiosyncrasies of the Tumbatu form of Swahili, *Kitumbatu*, however, were of interest to many, and were mostly attributed to people who actually lived on the island of Tumbatu, not all of those who identified as Watumbatu.

In my initial 18-month research period from January 1999 to July 2000, 76 cases were opened, and I looked closely at those 60 cases that were handled while I was regularly attending court sessions from June 1999 to July 2000. Of these 60, about 15 are presented in detail in this book and others will be mentioned in passing. Because of the short length of research trips in 2002 and 2005 (less than two months each), I was not able to follow any additional cases in full since few are resolved in less than two months. As a result, I do not include any complete cases from these years, but the interviews and court observations from these trips supplemented my previous work and allowed me to pursue questions that had come up after I finished the initial fieldwork. For the 60 cases I studied closely, I have essentially the same kind of ethnographic and documentary information, with minor variation. I have copious notes from nearly all the days the litigants were in court, and most court produced documents from each case. When I had to miss a day in court, I talked to Shaykh Hamid and the clerks about the day and made careful notes based on what they told me about what happened when I was gone. Also, although I managed over 40 formal interviews in the court with litigants (this is in addition to the casual conversation we might have about the case), this was certainly not all of the persons involved in the 60 cases. For a variety of reasons, it was far easier to formally interview plaintiffs than defendants. For one, most plaintiffs were female, and as a woman it was easier for me to talk to women than men about marital problems. (In and out of court, unless they were quite elderly, men were reticent to talk to me, an unmarried woman, about marital difficulties, and because I tried to adhere to norms of propriety, I did not push such conversations.) Also, in court, Shaykh Hamid and the clerks made a point of directing women litigants to me so we could talk after they had opened a case. They did not make a point of asking men to talk to me, but also did not discourage it, and I managed to conduct some in-court interviews with male plaintiffs. Furthermore, plaintiffs had a grievance they were usually willing to talk about, or else they would not be opening a case in the public space of the court. Most people opening cases were willing to talk to me, and no one ever refused when I asked. Defendants were much harder to interview, however, as some were angry about being summoned to court, and others seemed to move through their proceedings faster without lingering in the court; the *kadhi* and clerks did not direct them to me, and I was often uncomfortable asking them to talk.

Of the 40 or 50 marital disputes that are brought to the Mkokotoni court every year, maintenance claims, divorce suits, and pleas for the return of missing wives and restitution of marital rights are common. Of the 76 cases opened in the period under investigation, 70 concerned marital disputes of some kind. These numbers are fairly typical of the past 10 years in Mkokotoni: on average, about 40 cases are opened per year, and about 90 percent are marital disputes. Table 1.2 shows the number of cases opened per year by female and male plaintiffs between 1990 and 2004. The majority of cases in Zanzibar are opened by women; of all of the 76 cases opened in 1999–2000 period, women were plaintiffs in 61 and men were plaintiffs in 14; in the remaining case, described in chapter two, a woman brought a matter to court, but the case was opened as if her former husband was the plaintiff. This prevalence of women as plaintiffs is not unusual. Hirsch has shown

Table 1.2 All *Kadhi*'s Cases Opened in Mkokotoni Court, 1989–2004

Year	Total Cases	Female Plaintiff, Marital Dispute	Male Plaintiff, Marital Dispute	Inheritance, Child Custody, Other
1989	38	33	5	0
1990	46	35	11	0
1991	35	30	3	2
1992	46	41	4	1
1993	26	20	5	1
1994	23	17	5	1
1995	43	35	7	1
1996	48	40	8	0
1997	43	36	6	1
1998	51	37	10	4
1999	48	38	10	0
2000	46	38	7	1
2001	48	41	6	1
2002	34	30	3	1
2003	40	32	6	2
2004	41	34	6	1
	656	537	102	17
Total	656	(82%)	(15.5%)	(2.5%)

Note: The category "marital dispute" also includes those in which a woman sued for the right to be married.

that women were also more likely to open cases in Kenya's *kadhi's* courts (1998); Wurth gives numbers similar to mine from the 1988–1993 records from Yemeni court in Yemen (1995), as does Peletz for Malaysian Islamic courts (2002). Table 1.3 shows the number of cases involving disputes between husbands and wives that did not involve child custody; of these, women were plaintiffs in 57 and men in 13.

The largest category of cases opened in the Mkokotoni court involves maintenance claims opened by wives against husbands. Table 1.4 provides a breakdown of the type of cases opened; this table categorizes cases in the same way the court clerks listed case in the court register. It is clear that women open the vast majority of cases, and among those opened by women, a good portion involve a claim for improved maintenance from husbands. A number of cases involve women suing for divorce; those noted as women seeking registered divorce are those in which a woman sought a court document establishing the validity of a divorce that took place out of court. Those titled "pleas to be married" involved guardianship, and were opened by women suing for the right to be married to men of their choosing.

Most people in this part of Zanzibar refer to the Islamic courts as *korti* (court) *ya kadhi*. Very rarely, I heard the courts referred to as "love courts" or "women's court." Mzee Bweni, a friend and neighbor in his sixties who was the primary person who watched over me in Kinansi, jokingly referred to the courts as "the place of *chongoo*" (the one-eyed), which implied that the *kadhi* always decided in favor of women. Was Mzee Bweni right? Let's look at the outcomes of cases. In preparing this manuscript, I hesitated to prepare a table showing outcomes of the cases, or "wins" and "losses." As noted earlier, many researchers have noted that women tend to "win" cases in Islamic courts more often than they lose, or more often than men win cases in the same courts. In my work, I find it difficult to describe the outcome of a case as a simple "win" or "loss," even when the *kadhi* issues a clear ruling. For example,

Table 1.3 Marital Disputes between Husbands and Wives by Gender of Plaintiff, January 1, 1999–July 15, 2000

Type	Number
Marital disputes opened by wife	57 (81%)
Marital disputes opened by husband	13 (18%)
Total	70

Note: Table includes disputes between wives and husbands that did not involve child custody as primary complaint, N = 70.

Table 1.4 All *Kadhi*'s Cases Opened in the Mkokotoni Court by Type, January 1, 1999–July 15, 2000

Type of Case	Number	Details
Wife seeks maintenance	31 (41%)	In two, women asked for payment of *mahari*; in one, a woman claimed husband was impotent
Wife seeks divorce	13 (17%)	Various reasons specified, including lack of maintenance; abuse; impotence
Husband seeks wife's return or restoration of marriage rights	13 (17%)	Most involve men asking for a wife to return to the marital home
Woman seeks registration of divorce	7 (9%)	
Child Custody	3 (4%)	All were initially brought by women
Inability to get along	2 (3%)	Both opened by women against husbands
Daughter seeks permission to marry	2 (3%)	Both opened against father
Absent Husband	1 (1.3%)	
Woman sues former husband for pregnancy support	1 (1.3%)	
Abuse	1 (1.3%)	
Unspecified claim opened by woman against her husband	1 (1.3%)	
Inheritance	1 (1.3%)	

Note: N = 76.

if a woman sues for divorce from a husband based on his alleged failings in the marriage, but the *kadhi* rules that she should "buy" her divorce because of her own failings, is this a clear-cut win? Or, consider this: if a woman sues for divorce, and the *kadhi* rules that the couple must attempt reconciliation but notes that any failure to reconcile will result in divorce, is that a clear loss?

After some deliberation and at the suggestion of an anonymous reviewer, I decided that an outcomes table would be helpful. Table 1.5 is the result.[13] As it shows, the women "won" about 44 percent of cases they opened that ended with a ruling, and lost about 14 percent; 28 percent were settled by an agreement between husband and wife. In about six of these "wins," the *kadhi* ruled in favor of the female plaintiff because the husband never appeared in court. Men "won" about 53 percent of cases they opened, but lost far more than women, at

Table 1.5   Outcomes in the Mkokotoni Court, January 1, 1999–July 15, 2000

Plaintiff	Win	Loss	None	Settle
Women N = 57	25 (44%)	8 (14%)	7 (12%)	16 (28%)
Men N = 13	7 (53%)	4 (31%)	2 (15%)	0

Note: Table includes disputes between wives and husbands that did not involve child custody as primary complaint, N = 70. Columns "win" and "loss" refer to *kadhi's* rulings; "settle" refers to an agreement reached between husband and wife; "none" refers to cases in which litigants stopped coming to court. One case, which was the unspecified claim brought by a woman against her husband noted in Table 1.4, is not included here because I could not get any further details on the case.; this could also perhaps be included under "none."

31 percent. So, Mzee Bweni's view was not entirely accurate. Women, however, do lose fewer cases than men as plaintiffs, and as we will see in subsequent chapters, divorce is always available to women in one way or another, and many of the cases that were settled ended in divorce. However, often the ruling would please neither party, and it was sometimes difficult to determine who was more favored in the ruling. Even if a woman "won" a case by getting a ruling in her favor, it did not mean that she went home happy or got the outcome that she wished for. The same was certainly true for men. Also, as will become clear in later chapters, many cases did not end simply with a ruling from the *kadhi*, and the eventual outcome might have been very different. In the official court register, victors in cases were not marked: the columns simply indicated plaintiff, defendant, complaint, and *maelezo* (explanation), where clerks would record briefly what happened. For the sake of preparing the table, however, I determined that a woman or man was successful if the ruling rectified at least part of her or his initial claim. Thus, despite the obvious convenience of presenting this information in a table, I am not fully comfortable with representing case outcomes in this way, and I would ask readers to consider cases individually since the outcomes were more complex than what is indicated in the table.

Outline of Chapters

Chapter two looks at the daily workings of the Mkokotoni court. I introduce Shaykh Hamid and the other members of the court staff, and begin a discussion of the role of legal documents in court proceedings. Through an analysis of three court cases and the related court documents, the chapter focuses on the way in which disputes are framed by litigants, court staff, and *kadhi*.

In chapter three, I look at practices of marriage and divorce in rural Unguja. I also take up the question of what I call "disputed divorces." Many cases in the Mkokotoni court involve disputes about whether a divorce took place outside of the court. Most often, these are situations in which a woman believes she was divorced through unilateral repudiation but her husband denies that he divorced her. I propose that the prevalence of such cases cannot be attributed to gendered differences in legal knowledge, but from the fact that women and men have different experiences of divorce.

Chapter four also addresses disputed divorces by looking at how Shaykh Hamid determines intention (*nia*) of the actors involved in the disputes. In addition to a full analysis of the case of Abdulmalik and Mariam that was introduced in chapter one, the chapter also discusses three other cases in which intention was central. For Shaykh Hamid, knowing the facts of a case are not enough: he must determine the intention behind events because of the possibility of multiple meanings to particular divorce-related actions.

In chapter five, I explore the position of the *kadhi* between state and community through looking at the practice of witnessing. This chapter also examines how different types of authority figures are party to the dispute resolution process. I also consider how the *kadhi* differentiates between types of testimony and prioritizes the authority of particular witnesses.

In a number of cases, Shaykh Hamid justified that a decision that he considered technically unlawful was appropriate because it was fair or equitable. In chapter six, I consider the type of Islamic divorce known as *khuluu*, in which a woman compensates her husband financially. The chapter describes martial practice of making and planning for the gift of the marriage gift (*mahari*), and examines the financial matters involved in disputes that end up in *khuluu*.

CHAPTER TWO

Writing a Case: Court Actors and Court Procedure

I conducted most of my research in the court in the fishing village of Mkokotoni, which is pleasantly situated by a small bay. In Kiswahili, the word *mkokotoni* means "at the place of the mangroves," and indeed many mangroves grew in the shallow turquoise waters. Although Mkokotoni is not on the main road north, people from town often come to the market there because it is known for a fine selection of fresh fish. On any given day, the market vendors offered everything from octopus to skate to a delicious small fish known as *tassi*. A dirt road traveling northeast from the market area of Mkokotoni leads past a busy primary school to the whitewashed colonial-era building that houses several regional government offices. The *kadhi*'s courtroom is here, next door to the courtroom of the secular *jaji*, or judge. The location is quiet and breezy, away from the bustle of the marketplace, and the many verandahs of the building afforded a clear view of the small neighboring island of Tumbatu. An overloaded passenger boat that ferried people back and forth from Mkokotoni to Tumbatu was often in view, and when the tide was high, fishing boats like large dhows, *jahazi,* and tiny outriggers, *ng'alawa,* filled the small bay.

I rode to court on a new, though wobbly, bicycle every Monday through Thursday from my temporary home in the shady village of Kinanasi, about two miles away. Shaykh Hamid came to court via the crowded passenger boat from his home on Tumbatu, and we both arrived around 8 or 8:30 in the morning. The clerks always arrived a bit earlier. Although days in court varied greatly in terms of work-load, Shaykh Hamid usually left the court at about 12:30 or 1 in the

afternoon; it was only on very busy days that he stayed later. I often
left around the same time, and we would walk to the market together,
talking about our day, about my life in the United States, or his home
on Tumbatu. Shaykh Hamid would join friends for tea and a chat in
the open-air *hoteli* (small restaurant) before returning to Tumbatu, and
I would buy a cold soda and fresh fish to take home for lunch.

The courtrooms were located on the west end of the regional administration building (photo 2.1). This section of the building had its
own covered entrance, and people often waited there for their turn to
speak with the clerks or the *kadhi*. Most mornings, two or three people
were already waiting outside when I arrived, and we would greet each
other pleasantly while I parked my bicycle. One morning in July, right
after I started working at the court, I noticed a thin, tired looking
woman waiting on the steps with several small children. Her name was
Bi Amina. Like nearly all of the women who came to the court, she
wore a brightly colored *kanga* over her dress. A *kanga* consists of two
large rectangles of printed cotton cloth, usually adorned with a Swahili
proverb or saying or some sort; women wrap one piece around the waist

Photo 2.1   The Courtroom and Clerks.

and the other over the head and upper body. The fact that most women wear a *kanga* to court is indicative of the general poverty of the area. At the time, Zanzibari women with greater means nearly always wore a black *buibui* when going out in public. The *buibui* is subject to the winds of fashion and takes many forms, but is often a long, sometimes diaphanous, black cloak worn with a matching or contrasting headscarf. In the many months I spent working in the court, only a handful of women came to court wearing a *buibui*.[1]

Bi Amina had come to ask the court to require her former husband to take custody of their several children. She was in her forties, and recently experienced her third divorce from as many husbands. She had several children from that marriage. I learned that she lived only a 20-minute walk from my house, and I visited her several times over the next few months. Amina lived in a tiny two-room mud and thatch home that she called by the somewhat derogatory term *kibanda* (hut). She lived only with her children—she had no other family members nearby—and made an extremely modest living by farming a plot of land near her house. In court, she explained that she was very poor and did not have the means to support the children; what she grew and earned from selling produce was not enough to feed all of the children and pay their school fees. Although most people would agree that, according to *sheria* and local norms, the children's father had the duty to support his children, a case was not opened right away. In fact, it was only after many long days in court that the *kadhi* and clerks eventually agreed to open a case to address the problems. Working with a representative from the district commissioner's office, they decided it would be most effective to open the case as if Amina's former husband was suing for child custody. The claim document, which was the only official piece of paperwork produced for this case, specified the father of the children as plaintiff suing for custody of the children. Amina was listed as the defendant.

The remainder of this chapter examines the process and politics of creating court documents from courtroom activity. The historian Iris Agmon, who works on nineteenth-century Ottoman courts in Palestine, and has noted the dearth of studies of the way in which court documents are produced (1996, 2006; see also Borgene 2006). Like others, Agmon has identified the inherent challenges with using records from Ottoman courts, and comments that this is in part because of the clerks' selective process of recording. She recognizes that the events recorded in the *sijills* are not a reflection of what actually happened, and proposes that the documents may serve as a bridge of sorts "between the dispute and the *shari'a"* (1996: 132). What is necessarily absent from historical studies is a close look at the process of creating documents in a working court: how are clerks and judges deciding what information to include, and what to leave out? Although prepared documents and records reveal how the record-keeper, most often a clerk in Zanzibar's courts, understands a case, I have found that they can differ significantly from what happened in court, and from what the claimants or the *kadhi* understand as the essential legal issues of a dispute.

Brinkley Messick, an anthropologist who studies legal writing in Islamic contexts, takes a dialogic view of legal texts, and proposes the utility of a "textual ethnography" of court documents, which is one of the aims of this work. Building on the work of Bakhtin (1991), Messick argues that judges both create social acts and exercise power by highlighting particular issues. The processes of creating such texts are thus instances of "the production of power and the expression of authority" (1995). The cases under consideration in this chapter illustrate not only the "authorial control" of the *kadhi,* but also that of the clerks and, occasionally, other parties to the case. Historian Leslie Peirce has taken up similar issues in her work on a sixteenth-century Ottoman Islamic court, where she found noteworthy variation in how litigant testimony was recorded in documents (1998). She proposes that the difference in the way that spoken testimony was recorded was shaped both by the restrictions of procedure and the aim of preserving communal well-being; "The personnel, the legal dynamics and the communal locale of the court endowed it with a unique language, one that of necessity mediated between the needs of the law on the one hand and the needs of the court's users on the other" (270). As in Zanzibar courts, the way in which litigant testimony is recorded in the case documents was not arbitrary, but fundamental to the nature of the case. However, although documents demonstrate the way in which their preparers present legal issues in a formulaic way, they cannot clearly illustrate the variation in participants' perspectives and the different shapes a case may take throughout the proceedings.

This chapter adds an ethnographic dimension to work on disputes and documents in Islamic courts by exploring the often-circuitous route from testimony to text. Juxtaposing courtroom ethnography with court-produced documents illustrates the way in which the many different parties to a dispute may exercise authority and exert authorial control in constructing claims and resolving disputes. This process allows us to glimpse the particular legal issues court actors consider most important or, strategically, most persuasive. The clerks transform oral testimony and spoken claims into a written document, and the *kadhi* turns argument into a written account and uses it to inform his decision. Disputes are thus framed differently throughout the proceedings by litigants, by clerks, and by *kadhi*s, and the prepared documents are a manifestation of the entextualization of disputes.

The process of entextualization, or creating "extractable" texts from discourse, has garnered attention in the anthropology of language and linguistics (Bauman and Briggs 1990, Silverstein and Urban 1996, Hirsch 1998). In this chapter, I take up this issue through looking not just at the way in which litigants narrate troubles, but also at what the court staff do with these narratives. I aim to describe in detail the process of creating written texts from oral testimony, argument, and established procedure. Court documents may have many authors or contributors, each of whom may emphasize different legal issues at particular moments in the process and who build on previous entextualizations of the dispute. In a similar context, Susan Hirsch has shown that the Swahili litigants in Mombasa's *kadhi*'s courts "turn their stories into texts in gender-patterned ways" in an attempt to influence the way in which court personnel will interpret their narratives, and gives numerous examples of how women and men entextualize disputes in court (1998: 30). Although she notes that the disputants' narratives "are always potentially entextualizable by judges and clerks, who create literal texts of trouble stories in the records they keep" (29), her primary focus is on the way in which litigants present their troubles to the court rather than close consideration of how clerks and *kadhi*s interpret and record these narratives. I hope to contribute to this work by showing how clerks and *kadhi*s create these "literal texts." The remaining part of this chapter considers three disputes to examine the way in which spoken grievances and oral testimony are framed by clerks and *kadhi*s, who draw on their understandings of court procedure, Islamic law, and local practices of marriage and divorce.

The Court Staff

During my research periods in 1999–2000 and 2002, Shaykh Hamid Makame was the *kadhi* of Mkokotoni. He died suddenly in early 2005, and other *kadhi*s have worked in the court since then. When I returned to Zanzibar in 2005 and 2008, although things were running smoothly with the new *kadhi*, it was clear that the staff sorely missed the gentle Shaykh Hamid, and we always remembered him with great fondness.

For many years, Shaykh Hamid came to the court every Monday through Thursday. On a typical day, he would hear several disputes and talk with many disputants and visitors. On the rare days when no hearings were scheduled and few walk-ins sought his aid, Shaykh Hamid stayed in court to research cases, read relevant texts, and write decisions until midday. Most of the time, the staff at the Mkokotoni court consisted of one *kadhi*, two clerks, two messengers, and, toward the end of my first research trip, a typist. This was fairly typical of all of the rural *kadhi's* courts on Unguja. The court in Zanzibar Town, however, normally had three full-time *kadhis* and a much larger staff of clerks and typists. The Mkokotoni *kadhi's* courtroom was large, whitewashed, airy, and very sparsely furnished; the run-down fixtures and ramshackle furniture indicated that the room had seen better days, probably several decades earlier. At the far end of the room sat the *kadhi's* desk, which was large and covered in cracked leather. The desk was always tidy, and a well-worn leather-bound Qur'an and several aged books of *hadith* in Arabic were neatly arranged on top. In the center of the desk lay one or two current case files and an inexpensive ballpoint pen. Two wooden chairs for disputants sat facing the desk. The clerks' desk was near the doorway. It was much smaller and far less tidy; most of the time it was piled high with case files. I always sat on a chair against the wall near Shaykh Hamid's desk; I was often unlucky and got the worst chair in the room with the broken rattan seat. Until a typist was hired in the middle of 2000, the one very old and very noisy typewriter sat on the clerks' desk and the chief clerk, Bwana Fumu, did most of the typing by hunting and pecking with two fingers. Clearly, the courts lacked funding for new furnishings and updated equipment. As a thank you gift to the staff, I eventually ordered simple bookcases from a local furniture maker to shelve the case files, which were previously stacked on the floor.

Shaykh Hamid was in his early sixties when I met him for the first time, although he looked much younger. He was slim, with a delicately boyish, clean-shaven face. Like most *kadhis* on workdays, he wore a sparkling white *kanzu* (floor length shirt) over a *kikoi* (men's wrap skirt), a tweed sport-jacket, rubber thongs, and an embroidered cap called a *kofia*. He had a gentle manner, and out of courtesy and modesty, he normally refrained from prolonged eye contact with me and other women. On our first meeting, Shaykh Hamid seemed reserved and almost shy, but I soon realized that this was just his quiet, unassuming manner. Though authoritative in the courtroom, he was always patient rather than combative or dismissive with visitors, and it took quite a lot to provoke him to frustration or anger. Although Shaykh Hamid was not generally regarded as the premier Islamic scholar in the area, he was well liked by most people. We worked well together. He was always respectful to me and thoughtful, and never seemed to tire of my endless questions and requests for clarification on the details of particular cases. During my first few weeks in court, visitors were very surprised to see me sitting there with my laptop, and Shaykh Hamid always made a special point of introducing me as an American student of Islamic law, and kindly asked that everyone cooperate with me in my research. Eventually, word must have spread that an *mzungu* (a person of European descent) was working in the court, because even newcomers did not express much surprise at my presence. I never heard anyone complain or express concern that a foreign observer was in court.[2]

Zanzibari *kadhis* are appointed to the primary courts by the state through the Chief *Kadhi*, who (in theory) selects them based on their reputation as religious scholars. It was not possible to apply for the position of *kadhi*, and most *kadhis* I interviewed reported that they were surprised when suggested for the job. Shaykh Hamid explained that the Chief *Kadhi* nominated him for the position when the last *kadhi* of Mkokotoni fell ill and died. This predecessor had served as a *kadhi* since 1977, and Shaykh Hamid was *kadhi* from 1995 to 2005. Prior to his appointment, Shaykh Hamid taught religion for 22 years at a school on the island of Tumbatu. He was born on the island in the 1930s into a fairly typical family. His father worked as a fisherman, like many men in the area even today, and his mother farmed rice, cassava, and potatoes for a living. The *kadhi* told me that his father taught him to fish, and although had never gone to school himself, he had studied the Qur'an and *sheria,* which seemed to instill a love of learning in his son. Shaykh Hamid married twice and had several children, and until he died, he made his home on Tumbatu with his second wife, a tall and vivacious woman who was quite unlike her studious, soft-spoken husband.

Most *kadhis* of my acquaintance had traveled and studied abroad, but Shaykh Hamid did all his studies in Zanzibar and Tanzania. He and I discussed his life and religious trainings on numerous occasions, and during a field trip in 2002, he also prepared a brief written life history, and read it to me one slow day in court. I include most of it here because it was intended to assist me with my research.

I was born in the village of Tumbatu, Gomani on February 23, 1933. After that, in 1940, I began studying the Qur'an on the island of Tumbatu in a *chuo* (Qur'an school) led by Shaykh Muhiddin bin Hassan. I continued studying Qur'an until 1945, when I started to study the science of *fiqh* (Islamic jurisprudence), and matters of law—marriage, inheritance, and business.

At this, I asked the *kadhi* whether he had studied in a government school. He answered, and then moved back to his prepared history.

No, I did not... I continued there until 1954, then I joined the *chuo* that was led by Shaykh Tayyibu Hamdu bin Salim. This *chuo* was on the island of Tumbatu, and I continued to study the *tafsiri* (interpretation of the Qur'an) together with *tasawwuf* (Sufism)... matters of *zikri* (Ar. *dhikr*, Sufi practice to recollect God), of worship, of knowing Almighty God, what God does. Things like that.

I asked which Sufi order he followed, the Shadhiliyya or the Qadiriyya, both of which have a significant presence in Zanzibar (Purpura 1997).

I followed the Shadhiliyya in this *chuo* of Shaykh Tayyib. Then I continued until 1959, when I joined the *chuo* that was led by Muhammad Ayyubu in Tanga. We studied things like praise of the prophet.... And I learned to know the shaykhs there: Imam Shafi'i, Imam Hanbali, Imam Hanafi.[3] I stayed there until 1963, and then I joined a *chuo* that was led by Shaykh Muhammad Ramia in the city of Bagamoyo, [near] Dar es Salaam. And in this *chuo* I continued to study the science of *tawhid*—the science of oneness [of God]. This is the science of receiving God. [For example] Almighty God is the only one—there is not another one he does not have a brother, a child this is what we studied. I stayed there until 1973 when I was hired by the government of Zanzibar to teach religion in the school at Tumbatu-Gomani. I continued to teach there until 1995, when I was transferred, rather selected, to be the *kadhi* of the court of Zanzibar, to pursue work as the *kadhi* of the state at Mkokotoni.

At this point, I asked him to clarify his appointment, and wondered if he had requested the position:

Ataa! (Exclamation). No. I did an interview. And I didn't teach anymore because I was going to do this work. So from 1995 until today, I continue with this work.

Shaykh Hamid finished by telling me that he had summarized his progress in religious study, "In here [the document] were the *shaykh*s that I worked with, but they are all dead now. Now I remain, and I continue with this work of state-appointed *kadhi*."

Shaykh Hamid often expressed that he was pleased to be selected as a *kadhi* and he was very happy with the work. Occasionally, however, nominees for the position of *kadhi* turn it down. He told me that one of his own teachers was originally offered his position, but refused and recommend that Shaykh Hamid be considered.[4] *Kadhi*s and others scholars often explained that the nominees refused because they know they would not be able to apply Islamic law in full, since *kadhi* courts have jurisdiction only over family law. Occasionally, others hinted that *kadhi*s turned down the job because they did not support the politics of the ruling party in Zanzibar. Although *kadhi*s (and other judges) are not allowed to join political parties, friendship, political ties, and support for the ruling party are alleged by both lay people and some *kadhi*s to be a consideration in some *kadhi* appointments. Interestingly, although most of the *kadhi*s I asked were hesitant to voice support for a political party, it was evident through their discussion of current events that they differed greatly in their political views, and did not necessarily support the ruling party. Other individuals of religious learning explained that it was unseemly for men of religion to be involved in politics; this attitude is reminiscent of what Michael Lambek describes about Mayotte scholars in the neighboring Comoros as being "above politics" (1991: 34). Shaykh Hamid always claimed political neutrality and never expressed his political views to me, although he lived in area that was dominated by supporters of the opposition. He once joked that *kadhi*s "could be members of both parties or neither."

At the time, *kadhi*s received only minimal additional training after their appointment; this consisted primarily of rare seminars with the Chief *Kadhi* and other *kadhi*s.[5] Shaykh Hamid told me that the Chief *Kadhi* and the Mufti planned two yearly meetings for all the *kadhi*s to get together. The goal of the meetings was to further educate them in the law, and "correct mistakes" in their legal reasoning, and instruct them on how to use and apply *sheria*. He also noted that the Chief *Kadhi* had an advisory council that met every month to plan these biannual meetings. Although I do not recall Shaykh Hamid going to such a meeting during the time I worked with him, every month or two he went to the Chief *Kadhi* for assistance with a difficult case, and he had also studied a couple of short courses at the Islamic Call Society in Zanzibar Town. At the time of my research, *kadhi*s had no formal training in secular law. However, in a 2005 interview, an elderly *kadhi* who had recently been appointed to the court in Zanzibar Town told me that he thought it would be helpful if they received some training in secular law.

Although only the *kadhi* may write a ruling, *hukumu,* cases are undeniably shaped and thus in a sense created by the clerks, who open them, prepare written claims, and explain procedure and outcomes to the litigants. During all of my research, Bwana Fumu has been the head clerk at Mkokotoni. Bwana Fumu is a slight, quick (both mentally and physically) man who has an unassuming, almost bashful, manner. He is also an avid conversationalist who was always very curious about life in the United States. From the start, it was clear that Shaykh Hamid and Bwana Fumu had a close, mutually supportive relationship as colleagues in the court; the three of us enjoyed each other's company a great deal and spent many pleasant and, for me, productive hours together. When I met him for the first time, Bwana Fumu was about 40 years old and had worked at the court for about 10 years; as of 2008, he was still working there. He was thus intimately familiar with the proceedings, even though he had no formal legal training. Clerks are not required to have any special knowledge, and they are hired based on their personal qualifications for the job: in addition to being sharp and a quick study, they should have completed an adequate number of years of government schooling. Most have completed Form 4, which is approximately equivalent to Grade 12 in the United States. Without doubt, Bwana Fumu was Shaykh Hamid's foremost advisor and was consulted on every case for his opinion and advice. Court messengers and typists assist the clerks by fielding questions of newcomers, preparing and typing litigant's claims and responses, delivering summons and running general errands. It is possible to move between these positions, as evidenced by a confident young woman Bi Hamida, who was hired as a typist in mid-2000 and quickly promoted to the position of clerk within a couple of years. By 2005, she was working alongside Bwana Fumu as a primary clerk, and had become very comfortable working with and advising the new *kadhi* who had replaced Shaykh Hamid after his death.

In all of the primary courts on Unguja, the secular and religious judges are housed in the same building, and in all but the courts at Kariakoo and Mkokotoni, they use the same office and courtroom.[6] Clerks thus work for both the *kadhi*s and the secular judges. In the Mkokotoni court, although clerks technically worked for both justices, Bwana Fumu tended to spend more time working with the *kadhi* and another clerk worked more with the secular magistrates. In all of the courts, the clerks have specific duties and workloads. The clerk is the first to listen to the problems of claimants. He or she provides informal counseling and advises them to open a case or seek help elsewhere. If a case is opened, the clerks aid in the preparation of the plaintiff's official claim, *madai,* and the defendant's response, *majibu ya madai.* They also schedule court dates, manage case files, write summons, and explain procedure and rulings to litigants, their families, and witnesses. Bwana Fumu also organized case files for Shaykh Hamid and, because of his good memory for detail, frequently briefed him on the particulars of cases and the parties involved. Clerks are also required to record all cases in the court register, each entry of which lists the plaintiff, the defendant, the claim, and the *maelezo* (explanation); this final column is where a ruling or resolution would be noted. The entries are recorded in a very formulaic way down to the choice of words used. Clerks organize cases by type into a few categories: women requesting divorce; women requesting registered divorce; women claiming maintenance; men claiming their wives have left them; and the occasional requests for permission to marry, inheritance problems, or child custody problems.

Despite the absence of formal legal training, Bwana Fumu and the other clerks played an extremely important role in the workings of the Mkokotoni court. Bwana Fumu was the primary liaison between litigants and *kadhi.* As in other contexts, clerks reframe problems for presentation to the court and have a hand in determining the outcome. In Morocco, Rosen writes that "these clerks and attendant court personnel...often shape the assessment of 'facts' and channel the judicial process" (1995a: 199) and Hirsch observes that in Kenya "in their roles as record keepers, mediators, and *kadhis'* consultants, clerks wield considerable power to negotiate relations among the court, the state and the local community" (1998: 123).[7] In preparing claims, clerks frame cases in established legal categories and serve as liaison and interpreter between the spheres of the legal expert and nonexpert. In his work in Morocco, Léon Buskens has used the concept of the "culture broker," or those who "translate the events of everyday life into legally valid language" (2008: 144), to understand the significance of the role on professional witnesses who prepare documents. The concept is also particularly useful in the Zanzibari context because it is the clerks who shape initial litigant discourse, their tales of the "every day," into a court-appropriate written claim. The process of transforming the oral to the written not only represents what the documents' writers consider the essential legal issues of a case but also shows the strategy involved in presenting a dispute to the court. Litigants, clerks, and *kadhi*s have strategic reasons for framing disputes in particular ways. For example, a litigant may seek to avoid paying for a desired divorce or a clerk may present a claim in a way appropriate for the *kadhi*'s court instead of the criminal courts.

In her work on late nineteenth-century Ottoman courts in Palestine, Agmon considers the way in which court scribes create documents and how their record keeping contributed to the legal discourse on the family.[8] She suggests that it was possible that judges did not attend every session in a trial and scribes at times "conducted legal proceedings on the judge's behalf" (2006: 96). Because judges were often outsiders, the local knowledge of the scribes and clerks was key. Although some judges brought family members with them to work as scribes, other scribes and clerks were often long-standing local appointment, and in possession of the "local knowledge" that the judge lacked, and thus served as important liaisons between judge and local community (71). Furthermore, Agmon argues that scribes served as important liaisons between lay people, the "learned people of *shari'a* law," and legal notions of family. It was through the newly introduced process of reading court records aloud and requiring litigants' signatures that lay people interacted and gained an understanding of legal notions of the family (116, 123). As we will see, Shaykh Hamid did not rely on the clerks for local knowledge, since he was also a part of the community and was thus much aware of the norms people draw on to assess their situations and problems. However, clerks in Zanzibar similarly navigate between lay and professional understandings of the law. They essentially created the case and facilitated lay understanding of legal issues. It is the clerks who transform a marital grievance into viable legal claim that falls under the jurisdiction of the *kadhi*'s court.

Opening a Case

Procedure is consistent across the primary Islamic courts in Zanzibar, and in recent years attempts have been made to ensure that procedure in the *kadhi*s' courts mirrors that of the secular courts. The Mkokotoni court was staffed from Monday through Friday, and though Shaykh Hamid did not attend court or schedule hearings for Fridays, the clerks were there and able to assist visitors if necessary. When I worked there, the door was usually open, and people wandered in and out throughout the day. In addition to those seeking the aide of the *kadhi,* visitors included people from Mkokotoni, patrons of the nearby market, and workers from other offices who would drop by for a quick greeting or chat. Despite the seemingly informal atmosphere, the *kadhi* and staff maintained a strict protocol. Before opening a case, each potential litigant saw the clerk, who listened to the problem, asked a number of questions about the details, and then, if the problem required a court hearing, advised the seeker on whether he or she should open a case with the *kadhi* or with the secular magistrate. If the prospective defendant was not present, the clerk explained that a case must be opened before the defendant could be summoned; legally, an official summons cannot be sent without an open case.

Almost without fail, the first question that potential litigants were asked when they arrived at the court was whether they had been to their *sheha*, who has an official role in dispute resolution as a step preceding the courts.[9] If the seeker had already seen the *sheha*, then he or she was asked to produce the necessary letter. If not, the individual was instructed to see the *sheha*, explain the situation, and then come back with a letter. If a person was encouraged to open a case, the clerk explained that he or she must prepare the *madai* and pay a fee. When I began my research in 1999, the fee for opening a case was 1,150 shillings (about 1.75 USD at the time), and it has gone up a bit since then. The fee was a bit steep for many litigants, and it was sometimes waived if a claimant could not pay.

Although litigants were always told that they could prepare the *madai* themselves, I never saw or heard of anyone doing so, and it was thus always prepared by the clerks with the input of the claimant. It is in the preparation of this document that the clerks most clearly exercise their savvy and legal knowledge. To prepare the *madai,* the claimant is asked to review his or her problems for a clerk, who poses a number questions with the aim of isolating the essential legal problem. The clerk then writes the claim by hand in a specified format, beginning with the name, age, ethnicity, and home of the plaintiff and defendant, and numbering each subsequent point on a separate line. When a draft is ready, the clerk reads it aloud to the now-designated plaintiff, *mdai,* for his or her approval. If it was approved and no changes were necessary, the *madai* was typed in triplicate: one copy for the plaintiff, one for the defendant, and one for the court. Next, the *mdaiwa,* defendant, was summoned to appear. The summons letter was always hand-delivered by the court messenger to the defendant or to his or her *sheha,* who would see that the defendant received it. On his first visit to court, the defendant was presented with the *madai,* normally read aloud by one of the clerks, and instructed to prepare a response to it, called the *majibu ya madai*. As with the preparation of the *madai*, however, the *majibu* was always prepared by the clerks in the following manner: a clerk read the *madai* aloud line by line, and asked the defendant to answer each allegation by agreeing or disagreeing. The defendant was permitted to clarify points of the *madai* and to make a response, which were included in the document. The *mdai* and *majibu* are always written in Kiswahili.

During my research, the disputants' first audience with the *kadhi* was usually scheduled for the following week. However, because the clerks' desk was in the same room as the *kadhi's* desk, Shaykh Hamid often talked with the clerks and litigants during the opening of the case. In the years after Shaykh Hamid's death, however, a thin wall was put up to create a separate space for the *kadhi*, which undoubtedly prevented some of this initial interaction. Bwana Fumu told me that allowed for much more privacy in the discussion of cases. (Interestingly, during my many months sitting in court without a wall, people rarely complained about the lack of privacy.) At the first session with the *kadhi*, the plaintiff and defendant should both be present, but this was not always easy to achieve. People sometimes missed their scheduled appointments, and the steps of the court's veranda were often filled with those anxiously awaiting the arrival of a defendant or a witness. If one party did not appear, the session was rescheduled for another day. There was no formal legal representation in the *kadhi's* courts, and lawyers were never present.

Shaykh Hamid always prepared in advance for cases. As keepers of the court calendar, Bwana Fumu and the other clerks facilitated this by ensuring that the relevant case files were laid out on the *kadhi's* desk at the beginning of the workday. Thus, during the first session of a new case, Shaykh Hamid would have certainly read the *madai* and the *majibu*, and had perhaps already discussed the particulars of the case with Bwana Fumu. When the litigants arrived, Shaykh Hamid welcomed the party, and began by carefully explaining the goals of litigation and the required course of action in the courtroom. I noted that he explained the process exactly the same way in every case, and he took great care to make sure that everyone understood why they were in court and how the case would proceed because most litigants had not been to court before. After the introduction, the *mdai* and *mdaiwa* would each give oral testimony, and each would be permitted to question the other on the points made.

The plaintiff always gave her testimony first.[10] She was sworn in on the Qur'an by either Bwana Fumu or Shaykh Hamid, who would ask her to repeat a simple formula, through which she agreed to "speak the truth and only the truth." The defendant was instructed to remain silent while the plaintiff explained her side of the case. If a defendant was inclined to interrupt the testimony with questions, comments, or exclamations, Shaykh Hamid or Bwana Fumu politely reminded him to keep silent and to wait for his own turn to speak. Rather than asking pointed questions about the claims presented in the *madai* or *majibu,* Shaykh Hamid encouraged the litigants to tell their stories. He would listen quietly for long stretches, taking notes throughout, which would normally be included in the case file. Often, a narrative would go for quite a long time with little interruption from Shaykh Hamid other than requesting clarification on time, place, who was present, and so on. However, if a person proved particularly long-winded or had a tendency to ramble aimlessly, Shaykh Hamid would ask her to focus on the problems at hand. Like many other *kadhi*s, Shaykh Hamid was a teacher for many years, and this was apparent in the way he reviewed key points of the testimony in the manner instructing a student.

Next, Shaykh Hamid asked the defendant if he "understood" the plaintiff's testimony. If he had, the *kadhi* invited him ask the plaintiff three questions. It was not unusual, however, for a defendant to attempt to use this period to explain his side of the story. In this case, Shaykh Hamid would stop him with a reminder that there would be ample time for his story later; now was simply the time for questioning the plaintiff. If a defendant was puzzled, Shaykh Hamid would give examples of questions he might ask, like clarifications of the amount of food a husband provided. After the questions, the defendant was sworn in and asked to give his testimony, and then the plaintiff was invited to ask questions. Following this, Shaykh Hamid would decide if it was necessary to call *washahidi* (witnesses). If so, he told the litigants that they should bring someone who knew them well enough to be familiar with their problems. Witnesses were heard in much the same manner as the testimonies of the litigants: they were usually sworn in on the Qur'an and asked to share their knowledge of the litigants and their problems with the court.

At this point, each case proceeded individually. Some were closed within one or two hearings, and others took months depending on the number of witnesses called, availability of litigants, and circumstances of the problem at hand. Some cases were decided by the *kadhi,* others were settled by an agreement between litigants, and a few were left with no resolution; occasionally, litigants just stopped coming. When Shaykh Hamid decided a case, he produced a final document called the *hukumu* (ruling), which was often prepared in a format called *masharti* (terms of reconciliation). When a ruling was prepared, he called the entire party together so he or Bwana Fumu could read the document aloud and answer any questions. Furthermore, he always informed the litigants that if they did not like the court's decision, they had one month to appeal (*kukata rufaa*) to the Chief *Kadhi* in Zanzibar Town. It was unusual, though not unheard of, for litigants to appeal.

When I first worked with Shaykh Hamid from 1999 to 2000, *kadhi*s were not required to adhere to a particular format in writing decisions, and there was thus much variation in the length and content of written decisions from different *kadhi*s. The *kadhi*s also varied in the frequency with which they cited the Qur'an, the *hadith* literature or other legal texts to justify their decisions. Because he regarded most of his cases as simple matters, Shaykh Hamid only rarely cited legal texts in his decisions, which seems similar to what Dupret has noted about Egyptian judges (2007). When he did, it was usually the Qur'an or *hadith,* and he noted the Shafi'i scholar Nawawi's collection *Riyadh as-Saluheen* as a helpful source that he normally kept at hand. Although he did not always do so, Shaykh Hamid told me that, ideally, he should always cite the relevant *hadith* and Qur'an in case a decision was appealed to the Chief *Kadhi,* who would need to know how he had reached the decision. In complicated cases, Shaykh Hamid would also cite Shafi'i legal texts; one of texts he kept on his desk was the Shafi'i *fiqh* manual *Kifayat al-Akhyar.*[11] However, when we worked together, he included such citations in only two rulings. Shaykh Hamid always wrote his decisions by hand in Kiswahili. However, if he cited an Arabic text, he would do so in the original Arabic. He did not normally translate the Arabic into Kiswahili, but would usually note the source and the significance. Shortly after my first field trip, *kadhi*s were required to write decisions in a special format, and to cite relevant legal texts in every decision, not just those cases they deemed complicated or unusual. In 2002, Shaykh Hamid told me that some *kadhi*s were also writing decisions together to check each other's reasoning. Bwana Fumu explained that this was instituted so that decisions more closely resembled the formatting in the secular courts, as the *madai* and *majibu* already did.

Was Edda Broken? *Machano v. Aisha*

The first case we will consider was opened by a middle-aged man named Machano,[12] who came to court in early February 2000 with a claim against a woman called Aisha. Machano was tall and thin, and wore a men's woven wrap skirt called a *kikoi*, a ragged white shirt, and old rubber sandals. This was typical attire for men of his age and older; younger men usually wore trousers. He seemed confused when he arrived in court, but Bwana Fumu briskly told him to sit down in front of the clerks' desk and explain his problem. Machano sat down tiredly, showed Bwana Fumu that he had a letter from his *sheha*, and then told his story. He explained that he had divorced his wife Aisha, and shortly after decided that he wanted her back. He went to her family home to take her back, but she refused him and had married someone else. Bwana Fumu asked him how much time had passed between the divorce and when he went back to her. Machano, clearly upset, said that only two weeks had elapsed since the divorce, and argued that she could not possibly have maintained *edda* before remarrying. *Edda* (Ar. *'idda*) refers to the mandatory Islamic waiting period of several menstrual periods before a woman may remarry after divorce. The appropriate *edda* in this context was considered three periods, and thus because she had only waited two weeks, Machano determined that Aisha was unlawfully married to her new husband.

A debate ensued between the clerks about whether this was a criminal or civil case: if Aisha's guardians had allowed her to marry before she completed the waiting period, then it would be a criminal case of adultery because she would be with a man who was not a lawful husband. The clerks discussed how many days had passed since the divorce and whether a woman could possibly menstruate three times in that time frame. As the only woman in the courtroom at that moment, they consulted me on the matter, but I claimed ignorance since I was uncomfortable participating in the reasoning process. No consensus was reached, and eventually, Bwana Fumu asked Machano what he wanted to accomplish by coming to court. Machano replied simply, "I want my wife back."

At this, the clerk decided that he should open a case in the *kadhi*'s court since he wanted more than simply prosecuting Aisha for violating *edda*. When men bring cases to court, their most common claim is demanding the return of an absent wife and concomitant restoration of marital rights, and Bwana Fumu decided to frame Machano's *madai* as such. As usual, he prepared the document by asking Machano specific questions, then typing the claim in a standard format and reading it aloud for Machano's approval. Although the alleged violation of *edda* was not mentioned, the *madai* included Machano's claim that Aisha married another man while still his wife.

Madai

1. Plaintiff, male, age 45, Mtumbatu, from M—.[13]
2. Defendant, female, age 30, Mtumbatu, from C—.
3. They have been married for 20 years; have had 9 children, 2 of whom died.
4. The plaintiff claims that he divorced his wife on the twenty-fifth of the month of *mrisho* 1999 and he returned to her to take her back around the tenth of the month *mfunguo mosi* 2000; he went to her elder brother.[14]
5. The plaintiff claims that the defendant refused to return to him when he asked her to do so. By the twentieth of *mfunguo mosi*, the defendant was married to another husband.
6. The plaintiff demands that the defendant return to him because he is indeed her true husband.
7. This claim began with a complaint that the defendant married another man while she was the wife of the plaintiff.
8. The claim and the plaintiff originate from M—, Northern A district.
9. The plaintiff asks the court:
   a. To order the defendant to return to him immediately.
   b. To pay all relevant court fees.
   c. Follow any other obligations incurred.

Aisha was summoned and came to court with her new husband, Kassim, the following week. Although I had not met them before, I recognized the couple immediately. They lived down the road from me, and I had seen Kassim riding by on his bicycle on many occasions. Aisha was petite, and vivacious; although she was probably closer to age 40 than 30, she could have passed for a very young woman. Kassim was far older, perhaps around 60, but he had the same lively personality and quick smile. He walked with a distinct swagger and often had a cigarette tucked jauntily behind his ear; he could not have been more different from the dour Machano. He and Aisha appeared to get along well.

The clerks read the *madai* to Aisha, and proceeded to question her. Kassim, who always accompanied her and was an active participant in all of the court proceedings, informed the *kadhi* that although Aisha was lawfully divorced and had paid Machano for it, Machano refused to give her a paper indicating that she was divorced. The *kadhi* assured him that Machano would be required to produce the paper. Aisha asked how the case would proceed, and Shaykh Hamid told her that since she had given money to Machano, it seemed that her divorce was an instance of *khuluu*, in which a woman compensates her husband for a divorce she desires. Shaykh Hamid asked Aisha specifically about maintaining *edda* and when she had married Kassim. She answered that she had completed the waiting period, and that she married Kassim on the 19th of *mfunguo mosi,* which was about the same date that Machano had cited in his *madai.*

I interviewed Aisha that day in court, and she cheerfully shared details about her various marriages and her dispute with Machano. Kassim stood closely by listening carefully to our conversation; he was intrigued with the unfolding drama and added details here and there. Aisha told me that she had not had any formal education other than a bit of Qur'an school when she was a child. She had been married to Machano for many years, and over the course of their marriage he had divorced her twice. He asked her to return after the first divorce and she agreed. They stayed together until recently, when he divorced her for a second time. Aisha explained that when he asked her back after the second divorce he did not come to her family home, as he had said, but rather asked her to return when they met farming in the fields. When she refused to return to him, he asked her for money, and indicated that he would accept the return of her marriage gift, the *mahari.*

Together, Aisha and Kassim reemphasized that Machano divorced her on his own initiative, and then asked for money when she refused to return to him. They said that she paid him 50,000 Tanzanian shillings (about 45 USD at the time). Aisha said that this was more than her *mahari,* although she could not remember the exact amount. She told me she knew that Machano opened the case because he thought she had violated the *edda,* but she counted out the days and assured me that she had observed the full waiting period. She explained, however, that it was true that she had married as soon as possible, "Life is difficult for a woman without a husband, *sikai mjane mie!*" (I'm certainly not going to remain a divorcee!). Bwana Fumu prepared her *majibu ya madai* that day.

Majibu ya Madai

1. The defendant agrees with points 1, 2, and 3.
2. The defendant does not agree with the plaintiff's explanation in point 4. After the plaintiff divorced her, he did not return to her like he said. Rather, she remembers that one day she went to farm

her potatoes and there they met each other; the plaintiff then told her that he wanted her to return to him, "I told him that I would not return because I was tired of his behavior, and it is not true that he went to my elder brother because he is not at all nearby—he lives far away."

3. The defendant agrees with point 5 that she married another man on the nineteenth of *mfunguo mosi* but it was not like he claims: she married because she had already given the defendant the money that he wanted in the amount of 50,000 Tanzanian shillings; indeed, he received it, "I then decided to remarry another husband."

4. The defendant does not agree with point 6; she explains that she cannot return to the home of the plaintiff because he is not her husband and she is lawfully married to another man.

5. The defendant does not agree with point 7; she explains that she is no longer the wife of the plaintiff and she has been lawfully married to another man.

6. The defendant agrees with point 8.

7. The defendant asks the court:
   a. To throw out the claims of the plaintiff because he has no foundation on which to base these claims as she has lawfully married another man.
   b. To order the plaintiff to pay all court fees.
   c. For any other orders the court gives to reach an agreement.

Machano and Aisha came to court together some days later, when the *sheha* of their community was called to give his testimony. Because *sheha*s are expected to be familiar with the happenings of their communities, they are often called in by the *kadhi* as supplemental witnesses. The *sheha*'s statement in this case, however, did not indicate much knowledge of the situation. He said that he did not know anything about their problems: he had not heard that they had divorced or that Aisha had remarried.

Witnesses were called, and several days later the case resumed. Machano's witness was one of his older male relatives, Mzee Adamu, who explained in a loud, confident voice that he knew about the divorce and claimed that he himself had told Machano not to divorce Aisha and not to take her money. He said he knew of Aisha's new marriage, but when the *kadhi* asked for details, Mzee Adamu admitted that he did not know much about it or know who gave her permission to marry. Shaykh Hamid, mildly irked, turned to Machano and told him that he really should have brought a witness who knew something about the case.

At this, Aisha protested that the testimony was irrelevant because the witness did not know enough about their circumstances. Shaykh Hamid replied that the testimony—though certainly not ideal—was still useful because the witness knew about the divorce and knew that Machano wanted Aisha back. This comment was significant because it was the first time Shaykh Hamid emphasized the necessity of establishing the actual circumstances of the divorce. He kept the witness for some time, and asked more questions about the divorce event. The *kadhi* specifically tried to determine whether Machano had clearly expressed his desire to go back to his wife and when he had done so. He also wanted Mzee Adamu to confirm that Machano had asked Aisha for money and had received it from her. Mzee Adamu said that Machano had taken the money, although he repeated that he himself had advised him not to do so.

Aisha's witness was the local Qur'an teacher Mwalimu Simai, who was highly respected in the community for his religious knowledge. As was his usual practice when a respected elder came to court, Shaykh Hamid explained to Mwalimu Simai that he had been called in to help him "solve the couple's problem," rather than simply as a "witness." With less esteemed witnesses he simply asked them to explain their relationship to the couple and their knowledge of the problems at hand. Mwalimu Simai explained that although he himself had contracted the marriage between Aisha and Kassim, he knew only a little about the case. (After passing an exam, people like Mwalimu Simai can qualify to officiate marriages and prepare official marriage certificates.) When Shaykh Hamid asked him if they brought him an *alama* (sign) that they could lawfully be married, Mwalimu Simai answered that although they did not bring him anything, he believed Aisha when she said that she had observed *edda* and could remarry. Aisha interjected to explain once again that she had thought it prudent to remarry and so waited for the *edda* period and then married again immediately.

At this, Shaykh Hamid said to Machano, "You heard her: she says you divorced her, she waited, and then she got married again." Machano nodded his understanding, and the *kadhi* continued, "You said you divorced her, then you returned to her, then you asked for money, and then she gave it to you. Correct?" Machano agreed, and explained that he had received 50,000 shillings from her. After this, the session ended and the party left the court.

Shaykh Hamid decided the case the following week, which was early April, and thus two months from when Machano had opened the case; this was fairly typical. Many cases dragged on much longer, and a few were resolved more quickly. The *kadhi* ruled that Aisha had been validly divorced through *khuluu* and was legitimately remarried to her new husband. In the written ruling, Shaykh Hamid highlighted what he determined to be the essential matter of the case: the validity of the divorce action. He only briefly referenced the *edda* issue, and instead emphasized that Aisha had been validly divorced through *khuluu* and cited a verse from the Qur'an that established the conditions for such a divorce.

Hukumu

After listening to the plaintiff and the defendant and investigating their claims, the court has made a decision. The defendant has not erred because the plaintiff stated himself that she paid 50,000 shillings [for the divorce] and that he was given it. And indeed the plaintiff received it. He agreed that he divorced his wife. The court recognizes that he received it after he divorced his wife.

The days of *edda* were completed. Therefore the defendant, Bibi Aisha, is not the wife of the plaintiff Bwana Machano.

"If you fear that you cannot maintain the bounds fixed by God, there will be no blame on either if the woman redeems herself."[15]

And he divorced her by means of *khuluu* according to the laws of Islam. And the plaintiff does not have the right to harass his [former] wife because she is already married to another man. This case is closed.

Shaykh Hamid read the decision aloud to the litigants, and then proceeded to explain his reasoning. He told Machano that he could no longer bother Aisha with questions about her new marriage because it was lawful. However, he did have the right to appeal the decision. Aisha and Kassim were very pleased with the ruling, but Machano was plainly unhappy and bitterly tried to snatch a piece of Aisha's clothing as he ran by her when they left the courtroom.

In this case, ranges of different legal issues were emphasized by the litigants, clerks, and *kadhi*. Machano opened the case because he thought Aisha had violated the law by remarrying before completing *edda*. At first, he stressed this as the essential issue; it was only later at Bwana Fumu's query that he stated his desire to remarry. Bwana Fumu advised him to open a case in the *kadhi's* court rather than with the police because he deemed the important issue to be Machano's request for his wife back—not the potential violation of *edda* and alleged adultery. As we have seen, the *madai* stressed the return of Aisha as his wife. In his first hearing with Shaykh Hamid, Machano explained the *edda* problem even though the *madai* emphasized his desire for her return. The *kadhi* acknowledged his anxiety about the waiting period, but considered it settled with Mwalimu Simai's testimony: this religious authority had married Aisha and Kassim, which sufficed as evidence of her legal status to remarry—no other proof of a completed waiting period was requested or expected. Instead, Shaykh Hamid emphasized the validity of the divorce that had taken place, which he had ascertained through questioning the litigants and witnesses. Validity was established through determining that Machano had indeed tried to return to Aisha, and her refusal indicated her desire to be divorced from him. According to Shaykh Hamid, valid *khuluu* meant that a woman compensates her husband for a divorce she desires.[16]

They Have Both Made Many Mistakes: *Mosa v. Juma*

The next case was brought by Mosa, a middle-aged woman who came to court in October with a complaint that her husband, a man called Juma, did not provide adequately for herself and her children, some of whom were from a previous marriage. Mosa was a soft-spoken woman with a gentle manner and timid expression and, at first, she seemed somewhat hesitant to answer the clerks' questions about her marital problems. However, her daughter Fatuma, a stout and fearless woman of about 30, accompanied her and with her encouragement Mosa became more confident. When Bwana Fumu again invited her to explain the situation, Mosa told him that she was primarily concerned about maintenance: Juma failed to provide adequate food for the household or enough money for the children's school fees. After explaining these problems, she quietly suggested the possibility of filing for divorce. Fatuma was not pleased at this suggestion, for reasons which will become clear.

In the first of our several conversations about her case, Mosa said that she came to court because of maintenance problems, and she emphasized that she did not have a house to live in; she also told me that Juma refused to support the children from her first marriage. Juma was her second husband and they had four children together (one of whom died) and she had four additional children from her first marriage. She had no profitable work other than farming, and she said that Juma did not work or farm. Although she specifically told me that she had not asked Juma for a divorce, Juma had previously "written her for money." This phrase refers to a practice that is not uncommon in this part of Unguja, in which men ask wives for money to divorce them through unilateral repudiation. Most scholars and *kadhis* consider this practice reprehensible and unlawful because women are asked to "pay" for a divorce that their husbands initiate. We will look at this more closely in the next chapter.

In the initial stages of the court proceedings, Fatuma counseled her mother not to ask for a divorce, even though that is what Mosa ultimately wanted. This advice showed the younger woman's knowledge of the various types of Islamic divorce, and how the *kadhi* tended to use them. If a woman demonstrates that her husband failed in his marital duties, abused, or deserted her, she can be granted what is termed colloquially as a "free" divorce (*faskihi*, Ar. *faskh*). However, if a woman requests a divorce against the wishes of her husband, and cannot prove that her husband failed in some way, she may end up "buying" a divorce through *khuluu*. By advising her mother not to express her desire for a divorce, Fatuma sought to avoid *khuluu* by keeping the court focused on Juma's alleged failure to maintain her, which could result in a free divorce.

When Bwana Fumu prepared her *madai*, Mosa heeded Fatuma's advice by stressing that she had *not* asked for a divorce and instead emphasized Juma's failure to maintain her and the children; as Mir-Hosseini has observed among poor Moroccan women, Mosa and Fatuma were using maintenance as a route to divorce (2000). Bwana Fumu listened, and then asked if she and her husband got along. When she looked puzzled, he gave examples of the kinds of strife they might have. Did they use foul or abusive language? Did they argue? Were they rude or inconsiderate to one another? Mosa listened carefully and agreed that they did not get along very well and that Juma used unpleasant language, but she continued to emphasize that her main problem was the lack of a house and food, and Juma's refusal to support her children from a previous marriage. In the written *madai*, however, Bwana Fumu stressed instead what *he* considered the vital elements of her claim: the couple's inability to get along and Juma's bad language.[17] The document referred to maintenance problems only in a general way; Fumu wrote that one of Juma's "bad habits" was that he did not support her. Mosa's demands were written simply that he support her in the "normal manner."

Significantly, the *madai* did not mention that Juma did not support all of Mosa's children. As with Machano, Bwana Fumu framed the *madai* to represent the elements of her complaint that would make a reasonable case for the *kadhi's* court. Later that day, Bwana Fumu told me that a man is legally responsible only for his own children and that therefore Mosa did not have a strong maintenance claim, which is why he also emphasized the foul language.

When he finished writing, Bwana Fumu read the document aloud to the two women. Mosa approved and, because she had only little education, Bwana Fumu had to show her how to sign it. With his coaching, she signed her name painstakingly slowly in Roman letters.

Madai

1. Plaintiff, female, 44 years old, Shirazi, from K—.
2. Defendant, male, 50 years old, Mtumbatu, from K—.
3. The plaintiff and the defendant are man and wife and have been married for the past 18 years; they have 4 children, one of whom died.
4. The plaintiff claims that she does not get along with the defendant and he abuses her with bad language and does not fulfill lawful maintenance for her.
5. The plaintiff demands that the defendant stop his habits of not supporting her and of abusing her with foul language for no reason; if the defendant is unable to stop these bad habits then she demands he give her a divorce.
6. The plaintiff asks the respected court of the *kadhi* to listen to this claim stemming from the village of K in the Northern A district in the region of the northern Unguja where the court has the authority to listen to these claims.
7. The plaintiff asks the respected court to rule that the defendant follow the terms below:
   a. The court orders the defendant to support his wife in the normal manner of wife and husband.
   b. If the defendant fails to stop his habit of verbally abusing the plaintiff, then he must divorce her.
   c. The defendant should pay the court fees.

Juma was summoned to court the following week. He was a small, mild-mannered man. When Bwana Fumu read the *madai* aloud to him, Juma countered that he had done nothing wrong and that their marital problems stemmed from the fact that Mosa left him for no reason. His response was prepared with Bwana Fumu, and it stated that he was without guilt. In it, Juma claimed that he had not verbally abused Mosa, and that she blamed him for negligence to hide the fact that she had left him. Juma also claimed that Mosa left him without his permission or a reason, and that he wanted her back. When Bwana Fumu read the document aloud, Juma approved it, and then signed his name at the bottom in Arabic script.

Majibu ya Madai

1. Concerning points 1–3 of the claim, the defendant agrees with the plaintiff's explanation.
2. Concerning point 4, the defendant does not agree with the explanation and he answers that he is able to get along with the plaintiff although the plaintiff indeed left his home without his permission.
3. Concerning point 5, the defendant answers that he has not verbally abused his wife but the defendant gives this reason to hide her own fault of leaving her husband without his permission.
4. Concerning point 6, the defendant answers that he has no difficulties and asks that the court listen to his claim.
5. Concerning point 7a, b, and c the defendant does not agree with these requests and answers that the court should throw these demands out because, indeed, the plaintiff left the defendant without a reason.
6. Finally, the defendant asks the respected court to rule that the plaintiff follow the terms below:
   a. Return to the home of the defendant.
   b. The court fees are the responsibility of the plaintiff.
   c. Any other orders the court gives to reach an agreement with the defendant.

A week later, Mosa and Juma appeared in court together to give their oral testimonies. Mosa explained that when Juma asked her to marry him, she had told him that marriage might be imprudent because she already had children and he had another wife. They decided to marry anyway and Mosa received 7,000 shillings for her marriage gift, the *mahari*. She explained that she continued to live at her family home after the marriage because she was taking care of her children. Eventually, however, she moved in with Juma.

Shaykh Hamid asked about her marital problems and Mosa answered that she did not receive any clothing and that the food that Juma provided was not sufficient to feed all of her children. She added that Juma had "written her for money" for a divorce the previous year when she was in town visiting a relative. She told the *kadhi* that he had asked her for 70,000 shillings, an amount 10 times her *mahari,* and that she did not understand why he thought she had so much money. She did not pay him, and they eventually reconciled and lived together quietly for the next year. When Shaykh Hamid asked her if there were other problems, she said that Juma had promised to build a new house for her but had not yet done so. In his testimony, Juma explained that the troubles began when Mosa left him without an explanation. He said that eventually he went to her elder brother and found out that she left him because his house was *mbovu* (rotten), and in a poor state of repair. Juma argued that he had problems with her children, and declared that they were all disrespectful and foul-mouthed, and that her daughters brought their lovers back to the house. When the *kadhi* asked him about the 70,000 shillings he requested, Juma replied that he had indeed written for money because she was away for such a long time, but that now he no longer desired a divorce.

They were back the following week with witnesses. Juma's witness was a male relative who testified that he knew that Juma had "written for money" and that he had had problems with Mosa's children. Unsurprisingly, Mosa's witness was her daughter Fatuma. In her testimony, Fatuma vehemently supported everything her mother had said, and reemphasized that Juma did not support her mother properly.

Shaykh Hamid decided the case the following week. He told Mosa that both she and Juma had made mistakes in the marriage, and ruled that because they had lived together peacefully until they brought their problems to court, they must try living together again and upholding their respective marital duties. His written judgment was in the form of a contract, *masharti,* that each must follow, which was typical when the *kadhi* thought a couple capable of reconciliation. Juma was not ordered to support Mosa's children from her previous marriage because, as Bwana Fumu had explained, he was responsible only for his own children. It is also worthy of note that the *kadhi* ordered the litigants to report any further problems to their *sheha.* Bwana Fumu read the decision aloud, and the *kadhi* questioned the litigants to determine whether they understood what was expected of them. Juma looked at the floor throughout, and Mosa looked straight at the *kadhi.* Both said that they understood.

Hukumu

Terms for the Plaintiff

The plaintiff has children who have been disrespectful to their elders, therefore the plaintiff must teach her children good language and manners. She should share her life with her husband and leave behind her harsh words and behavior. She must respect the marriage because it is an order of God. She must follow these orders of marriage because if she does not she will break the law of Islam. If she makes any mistakes, then she must buy her divorce from her husband.

Terms for the Defendant

The defendant must recognize that he is the provider for his wife. He is expected to get along with her and use good language, to live well with her children, to support her with food, clothing, and a house, and he must give her money for household necessities such as hair oil, perfume, etc. He must fulfill all these terms and if he does not, he must divorce his wife if she brings in witnesses and a letter from the *sheha* attesting to his misdeeds.

Terms for Both

Both the plaintiff and the defendant must give a report of misdeeds to the *sheha*.

After about three months, Juma came back to court to claim that Mosa was not fulfilling the *masharti*. He brought a letter from the *sheha* stating that he had tried to live by the agreement but that Mosa was not upholding her end. Two days later, Mosa came in to claim that Juma only brought food in the afternoons, that he did not give her soap, oil, or money for the children's schooling, and that the house was still not adequate. The *kadhi* listened, but reminded her that she must first go to the *sheha* if she had marital problems, as was set out in the ruling. Frustrated, she left, and on her way out the door, she told me that she was reluctant to go to the *sheha* because she was an adult who already had great-grandchildren: "I'm embarrassed to go to the *sheha* every day to complain about my husband!"

One week later, Mosa and Juma were back in court together. Shaykh Hamid asked Mosa if she had fulfilled the *masharti*. She said that she had. He asked her why, then, she was not living with Juma as ordered. She answered, "Because he has no house!" The *kadhi* was surprised, but told her once again to go to the *sheha* to report the problem. When he asked Juma about the house, the man said that he did indeed have a house, but that she would not stay there because it was "a bad one." Mosa then argued that Juma was not giving her adequate food. Juma responded that he brought food regularly, but agreed that it was not enough for all of her children.

After hearing their complaints, Shaykh Hamid reminded the litigants that they had "two laws" to fulfill: maintaining the terms set by the court and reporting problems to the *sheha*. He explained yet again that when the court issues *masharti*, disputants must go the *sheha* if the other party breaks the terms. Failure to do so was a violation of court procedure. Thus, the law had been broken by violating the *masharti* and neglecting the *sheha*. Shaykh Hamid explained that Mosa would have been able to easily end her marriage if Juma had not fulfilled his obligations. However, both parties had made mistakes: they "were not sleeping in the same place," Mosa had not returned to Juma, and she did not go to the *sheha* to report the problem. Shaykh Hamid decided that they must try to reconcile one more time, and if Mosa could not bring herself to stay with Juma, she would buy her divorce. At this, he sent them home.

They were back about two weeks later, when Juma said Mosa would still not live with him. Shaykh Hamid immediately ordered Mosa to buy her divorce. She agreed without argument.[18] In the final ruling, in late March, Shaykh Hamid explained that Mosa agreed to buy her divorce in *khuluu* because she did not want to return to Juma. She was required to do so not only because she bore some responsibility for the marital discord, but also because she violated procedure by failing to see the *sheha*. Interestingly, although he took time to explain this to the litigants, he did not specify it in the written ruling.

March 28, 2000. In the court of the *kadhi* of Mkokotoni, the defendant says that from the day they were given the court's contract until today, "My wife has not agreed to follow me to my home and every day I go to her home to call her but she does not want to come. She was also called by the *sheha* but she did not go. This is indeed my explanation."

The plaintiff says, "I will not go to his home because I do not want him and he has no place to keep me and I will follow the terms of our agreement by buying my divorce from him according to the law."

The defendant agrees. The plaintiff requested a delay to bring the money to buy her divorce. After the agreement of the plaintiff and the defendant, the court read the *hadith* of the Prophet Mohamedi concerning *khuluu* divorce. And the *aya* of the Qur'an, which explains the process of *khuluu*. After this, the court wants the plaintiff to pay the amount with which she was married [*maharí*]; 25,000 shillings. She has until April 3 to bring it to the court.

April 3, 2000. The plaintiff and the defendant have arrived at the court. The plaintiff paid 25,000 shillings to buy her divorce from the husband. The defendant has received the money, divorced his wife, and the divorce paper has been written. This case is closed, and the plaintiff and defendant are now divorced.

When we talked about the case together, it was evident that Shaykh Hamid ruled in this way because he considered Mosa at fault in their marital problems, although his written ruling notes that both litigants had complaints about the other. The court never established that he was not supporting Mosa or their children adequately, but Shaykh Hamid included Mosa's statement that he did not have a proper house for her. Mosa's original claim that he was not supporting the children was not mentioned at all.

Amina's Children

Let us now return to the unusual case that was introduced at the beginning of this chapter. Recall that Amina had come to court because she was having great difficulty supporting her many children. She wanted the court to require her most recent husband, Seifu, who unfortunately had a reputation for laziness, to take custody of them and support them, as was his duty under *sheria*. Amina argued that the children should live with their father because there was no way she could feed and clothe them and pay their school fees on the meager income her farming provided. When she came to court the first time, the clerks and *kadhi* listened to her complaint together, but after some discussion determined that the matter should be taken to the police. Shaykh Hamid explained that they could not open a case for her in his court because her husband had the legal right to the custody of all of the children who were over age seven. Therefore, there was no real dispute between husband and wife over custody: Amina wanted Seifu to take the children, and the law gave him custody. If he and Amina both wanted custody, then there would be a case to hear, but in this situation, it was merely a matter of getting the man to take his children. Amina left somewhat despondently, as if she knew already that the police would not resolve the matter, and she did not come back to the court for several weeks.

I visited Amina at her home a couple of times with my research assistant Ahmada, and I learned that after the *kadhi* sent her to the police department, they had sent her to the *sheha,* who then sent her to the District Commissioner's office, which was in a village about 10 miles away. Although it was not clear to me why she was sent to the DC's office, the officer and his staff listened to her problems, and eventually determined that the best way to handle the situation was to ask Seifu to take custody of the children with the assistance of his older sister, if she would agree to help raise them. They came to this conclusion because of the man's reputation for idleness and unreliability, and they doubted that he could manage the children on his own. The District Commissioner then told her to take the matter back to the *kadhi* with their suggestions, and a formidable staff-member called Bi Mwajuma agreed to accompany her.

When they came to court and explained what they had discussed in the DC's office, the *kadhi* and clerks deliberated about whether Amina should open a case or seek an informal resolution to the problem along the lines Bi Mwajuma suggested. Bwana Fumu recommended the informal resolution, but the *kadhi* was leaning toward opening a case, most likely so that Amina would have a ruling to fall back on in case the arrangements went awry. He was still concerned, however, about whether he could actually open a case, and he explained again that "There is no [legal] case to transfer the children from the mother to the father since he has custody rights. Therefore," he said, "the father should just take the children!"

Nevertheless, the group finally managed to work out a solution: they should open a case as if Seifu was suing Amina for custody of the children. In most cases, as we have seen, the clerks prepare the claim document. This time, however, the *kadhi* did so with the assistance of Bi Mwajuma and the clerks. While he was writing the document, Shaykh Hamid asked Mwajuma what he should include in the *madai,* and did as she instructed. Most notably, she told him to include the father's sister, Safia, as a co-plaintiff.

Madai

1. The plaintiff is a man, aged 45, from K——.
2. The (second) plaintiff is a woman, age 49, from P——.
3. The defendant is a woman aged 40, from K——.

4. The first and second plaintiffs are a man and his sister who were born of one father and one mother.

5. The first plaintiff and the defendant were married and divorced and had six children together; one of whom died. [Children's names are listed.]

6. The first plaintiff and the defendant divorced one year ago, and all the children are with the defendant.

7. The first plaintiff and the second plaintiff together ask to take the care of these children from the defendant.

8. The core of this plaint is that the children will be taken in the custody of the plaintiff if he is indeed the father of the children, because they have reached the age at which they should live with the father according to the law.

9. The plaintiffs ask that the esteemed court of the *kadhi* listen to the request that arises from here in K—.

10. The plaintiffs beg the court to issue a decision that the defendant follow the following:
    a. The defendant must give up care of the five children according to the plaint.
    b. The court fees must be paid by the defendant
    c. Any other orders that come from the court and agreed upon by the plaintiffs.

A few days later, the entire party was present, including Seifu and his sister Safia. Seifu swore an oath that he was indeed father of the children and that he would take them and support them. Safia agreed to help raise the children. In his final notes on the case, the *kadhi* recorded that Amina stated that she agreed with the arrangements. I ran into Safia on my way home that day, who told me that the children were going to their father's home that very day. However, when I visited Amina the next day, she said that the children had already returned to her because they did not get along with their father's new wife. A couple of months later, however, the family had worked out a suitable arrangement: the children still slept at their mother's house, but went to their father's house every day for meals. Amina was content.

This case shows how the *kadhi*, with the noteworthy and unusual assistance of a representative from the District Commissioner's office, reframed Amina's request that her ex-husband support their children by opening the case as if her husband was filing for custody of the children with the aide of his sister. Although the *madai* was the only document prepared for the case, it shows how the actual progression of events in a particular case may not be mirrored in the court document. Furthermore, we see the willingness on the part of the court staff, most notably the *kadhi* here, to incorporate outside opinions and suggestions into courtroom deliberations and in preparing written documents.

Discussion

It goes without saying that court documents do not tell the whole story. Although a study of documents like the *madai* and *majibu* shows how the clerks present legal issues in a formulaic way, and the *hukumu* illustrates the way in which a judge writes rulings, court ethnography reveals the legal understandings different parties bring to a case and sheds light on the strategies of representation involved in the creation of court documents. Studying the process of entextualization in the three cases examined in this chapter shows differences in what court actors consider the essential legal issues in the disputes, and how they incorporate their understandings of law and local into the court and into the process of handling disputes. In studying the way in which women brought claims to Ottoman courts in nineteenth-century Palestine, Agmon asks, "were women really so knowledgeable and resourceful, or did they appear in court with a jumbled array of grievances and arguments that the judge then organized in order to extract logical *shari'a* claims, thus performing something of a legal service for them?" (1996: 132). She proposes that the courts played a powerful role in women's understandings of their marital rights and in framing their claims for the court. With the benefit of ethnographic research, we see that this is precisely what is happening we see in present-day Zanzibar courts. While male and female litigants do come to court with varying levels of awareness of the legal issues involved in their problems and with specific strategies at hand, it is the clerks who transform their "jumbled grievances" into a court-worthy claim.

With all parties, litigants, clerks, and *kadhi,* we see strategic framing of disputes to achieve a particular outcome. In the process of framing, litigants and court personnel draw variously on their understandings of Islamic law, norms of courtroom proceedings, and their understanding of local practices of marriage and divorce. Litigants may emphasize particular issues when they come to court that will be dismissed as irrelevant by clerks when preparing their claims. Buskens has explored the relationship between legal writing and daily life, and proposes that people like professional witnesses in Morocco are engaged in a process of "cultural translation" as they prepare documents (2008: 157). This is certainly what we see with the clerks in Zanzibar courts. Machano brought forth a claim that his wife had violated *edda* and had unlawfully remarried, and Mosa and Fatuma made a special point of framing Mosa's case as a request for maintenance rather than a request for divorce in an effort to avoid "buying" the divorce. Clerks, acting as "culture brokers" in the way Buskens describes professional witnesses, then reframed litigant narratives into a case appropriate for the *kadhi's* court and in a way that might enable the plaintiff to succeed in court. In Machano's case, Bwana Fumu determined that Machano wanted his wife more than he wanted her to be charged with violating *edda* and committing adultery, which would have been a matter for the criminal courts. Hence, the clerk instructed him to open the case in the *kadhi's* court and subsequently framed it as a familiar "missing wife" problem that would be appropriate for hearing in that court. In Mosa's case, Bwana Fumu also framed the claim differently from the way she presented it: he knew the *kadhi* would not require her husband Juma to support children who were not his own, and so emphasized Juma's bad behavior and foul language in the *madai*.

Though he does not consider specifically legal or religious texts, John Haviland's work on the entextualization of "texts from talk" among Tzotzil speakers of Chiapas facilitates an understanding of the process of the clerks' role in preparing documents. Haviland shows that the canons of writing are used as instruments of power in that they "have the power to define what counts note merely as correct, but also as 'sensible,' 'logical,' 'coherent' or even, simply, 'tellable'" (1996: 45). As in courts everywhere, Bwana Fumu has knowledge of state statutes, procedural law, and legal norms that the litigant does not, and therefore has the power to define what elements of narratives of marital discord should and will be emphasized. Often, Bwana Fumu used his ability to frame problems in appropriate legal language and enhance the claim of the plaintiff, whether male or female, in a way that bettered his or her chance of receiving a favorable decision from the *kadhi*.

The case is framed again when it is heard by the *kadhi*, and different issues may emerge as most salient. In the cases examined here, Shaykh Hamid's assessment of the legality of certain acts or the relevance of certain parts of litigant narratives often differed from the views of clerks and litigants. Agmon, like Messick, argues that the practice of legal writing is a discursive one, and she demonstrates how judges are at a crossroads of different legal ideas and influences (2006). Shaykh Hamid referenced not only the *madai* and *majibu,* but also the new narratives that are presented when the litigants and witnesses give testimony. Furthermore, the *kadhi* referenced not only court procedure, local practice, and modes of authority, but also Islamic legal sources of which the clerks litigants might not have expert knowledge. As a result, he regularly made a point of questioning the legality of not only the defendant's actions but also the plaintiff's. In Mosa's case, Shaykh Hamid questioned her own behavior in the recent history of the marriage, and eventually determined that Mosa was at least as much at fault in the marital discord as her husband, which resulted in her "buying" a divorce. In Machano's case, he made a point of determining whether Machano had divorced Aisha through proper *khuluu* or through unlawful "writing for money." Amina's case was a bit different, of course, but is illuminating because it shows just how vastly different court documents can be from what "actually" happened in the courtroom. Here, Shaykh Hamid and clerks worked with the representative from the district commissioner's office to create an appropriate "case" for the *kadhi*'s court by turning the woman who brought the claim into the defendant, as if she were being sued by her children's father for custody. Also, although it is not evident in a simple reading of the *madai,* the document was prepared with the authorial input of many.

CHAPTER THREE

From Community to Court: Gendered Experience of Divorce

One January morning, a gregarious woman named Shindano arrived at the court to ask for a receipt verifying that her husband, Abu Bakr, had divorced her. Shindano appeared to be about 60, wore a colorful *kanga,* and was barefoot. After waiting outside the courtroom on the steps for a short time, the clerks called her inside and asked her why she had come to court. Shindano told them about her husband's many violations of his marital duties and the distress he had caused her. She explained that Abu Bakr had divorced her, but later came to her home and announced that they were still married and demanded that she return to him: "He chased me out of his house, he got rid of all my *vyombo* (household goods) that were in the house and put them outside, and now he says that he didn't divorce me!" The clerks asked her for Abu Bakr's written statement of repudiation as proof of divorce. When she said that she did not have such a paper, they told her to bring her husband to court. When he came to court, Abu Bakr said that he had not divorced her and had never intended to do so.

A few months earlier, a young woman called Zaynab opened a similar case. She and her father came to court claiming that she was divorced because her husband, Rashidi, had told her to leave his home. She left, but when she went back to ask him for the divorce paper, he refused. She explained that Rashidi told her once again to go home and added insult to injury by calling her a "dog." When Rashidi came to court, he denied that he had divorced her. At about the same time Zaynab was in court, another young woman, Jabu, also came in with a dispute about whether she had been repudiated by her husband. Jabu had been married twice, and came to court because her first husband, Rajabu, whom she had not seen in years, had recently come to the home of her second husband to demand that she return to him. Like Shindano and Zaynab, Jabu claimed that Rajabu had divorced her: he had sent her out of his home years ago and told her to return to her parents. Like Abu Bakr and Rashidi, Rajabu came to court and said that he never divorced her. Furthermore, he argued that Jabu had remarried unlawfully, and was therefore committing adultery with her second husband.

Disputes in the Mkokotoni court often involve a concern about whether a divorce by repudiation, *talaka*, has taken place. Of the 70 marital dispute cases opened in the court between January 1999 and July 2000, 25 (or 34 percent of the total cases) involved a dispute about whether a man had validly divorced his wife by repudiation. In this period, seven women opened cases seeking a registration of an out-of-court divorce (referred to as *rijista, passi,* or *cheti cha talaka*); Shindano and Zaynab were among them.[1] The other cases were opened with a different initial claim, as with Jabu, but essentially revolved around alleged divorce. When considering disputes such as these, it might appear that the frequency of disputes about alleged divorce results from gender differences in legal and religious knowledge among rural Zanzibaris. Do women and men have different views of divorce procedure? Do women lack understanding of what a lawful Islamic divorce entails?

When considering earlier literature on the Swahili coast, one might be led to answer "yes" to both of these questions. Many scholars have addressed the subject of religious and customary knowledge among Swahili men and women. For example, Carol Eastman (1984, 1988) and Margaret Strobel (1979) suggested that religious law, *sheria za dini*, was the provenance of men in coastal Kenya, and that *mila*, an ambiguous term often translated as "custom," was the domain of women. More recently, however, scholars have reevaluated the *dini-mila* dichotomy. Middleton has criticized the association of women with *mila* and men with *sheria*, arguing rather that *mila* is "part of permitted Swahili religious practice, and efforts to regard it as forming part of a distinct female subculture are unfounded: both men and women accept and practice it" (Middleton 1992: 119). He proposed that earlier distinctions were problematic because they aimed to demarcate types of knowledge and access to modes of learning that, in reality, are less precise. Purpura made similar observations in her work among religious scholars in Zanzibar Town (1997) and Patricia Caplan noted that people of Mafia Island, south of Zanzibar, do not distinguish between *mila* and *sheria* along gendered lines (1995). Most recently, the contributors to the volume entitled *The Global Worlds of the Swahili* describe the ambiguity and complexity surrounding the terms *mila* and *dini*. The editors argue that it is misleading to view *dini* and *mila* as "opposite poles" but emphasize that the two overlap and intertwine, and note that they are used and understood differently in different Swahili contexts (Loimeier and Seesemann 2006: 10). In rural Unguja, I found it difficult to pin down one definition or understanding of the term *mila;* most often, people would use the term to indicate "culture" or "custom," though I also heard references to *sheria za mila* (the laws of *mila*). I found no indication that people think of *mila* as the domain of women and *sheria* that of men. Accordingly, women and men both seem to know what a "proper" Islamic repudiation entails. Even in rural areas, education is highly valued and girls and boys attend government and Qur'an schools at about the same rate (Montresor et al. 2001). Younger women are as likely as men to pursue the study of religion into their adult years, and women of all ages expressed an interest in increasing their knowledge of religion and religious law. Many reported that the radio was a primary source of information regarding religion and law.[2]

Given this, gendered spheres of knowledge cannot likely account for the prevalence of disputes over repudiation. However, although men and women do not appear to have different understandings of what religiously appropriate divorce entails, they experience divorce differently. Consequently, women and men highlight different events when describing their own divorces. A man's wife need not be present at the moment of repudiation for it to be legally valid, and it is not unusual for a man to divorce a woman when out of her presence. When we discussed the matter, the clerk Bi Hamida agreed that many cases involved disputed divorce, and said that women often assumed they were divorced without seeing or hearing a repudiation statement. Because of this, a woman may interpret other specific structural events as indicative of divorce, regardless of whether she has seen or heard the words of repudiation. However, it is noteworthy that when asked what a lawful Islamic divorce by repudiation entails, women and men give similar answers: a man must either write or speak the divorce statement. Moreover, when women take such issues to court, they do not claim that these structural events constitute a lawful Islamic divorce, but rather attempt to legitimize the experience through obtaining a receipt of a registered divorce.

This chapter examines how women experience and talk about of out-of-court divorces, and what happens when cases like those of Shindano, Jabu, and Zaynab are brought to court. In her study of disputing in Kenyan

Islamic courts, Susan Hirsch finds that although both men and women make use of several legal discourses, which include Islam and Swahili ethics, men are more likely to begin with Islamic law, and women more likely to move to religious law after utilizing other discourses of marital norms and marital disputing (1998). As a result, Hirsch argues that the discourse of Islamic law is "masculine" and men—not women—are therefore viewed as the proprietors of Islamic law (111). Other discourses also remain gendered, even though both men and women may use them. Like their Kenyan sisters, Zanzibari women draw on local norms when describing divorce both in and out of the courts. However, this is not because they lack knowledge of *sheria* but rather because they experience structural events of divorce directly, but may not witness the repudiation. As we shall see, Shaykh Hamid did not validate alleged divorces without proof of repudiation. However, he did not dismiss the cases as simple misunderstandings. Rather, he stressed the preservation of the marriage bond through reframing the cases as disputes about marital rights and spousal obligations under Islamic law. In such cases, most women did not press the validity of the alleged divorce. Rather, they acknowledged the *kadhi's* move to a discourse of rights and duties, and used this discourse to request either better maintenance or a court-ordered divorce.

Marriage and Divorce in Rural Unguja

The western portion of the district the Mkokotoni court serves is densely populated, and the villages along the one major road tend to run into one another; it is difficult for a newcomer to discern where one ends and the next begins, though residents have a clear idea of which households are in which villages. Kinanasi, where I lived, was one of about 10 villages in the *shehia* of Mnazi Mrefu. The population of the *shehia* was about 3,000 at the time of my research and *shehias* in the area ranged from approximately 2,500–6,000 people. People in this part of Unguja subsist primarily by farming, and arrangements are somewhat flexible. Both men and women are active in farming tasks, and husbands and wives may farm together, or married people may continue to farm with parents or other family members.[3] Married women often maintain economic ties to their own families, and I know several mothers and daughters who continued to farm together after a daughter's marriage. As we will see in later chapters, a woman's family is a critical financial resource in the event that she must "buy" a divorce. A wide variety of crops are grown, including cassava, bananas, rice, potatoes, coconuts, mangoes, among many others. Dietary staples include cassava, rice, and green bananas.[4] Farming is supplemented in a variety of ways. Many people are shopkeepers or participate in the informal economy. Women often sell homemade baked goods, juice, home-pressed coconut oil, or woven baskets and mats. This part of Unguja, particularly the village of Mkokotoni and island of Tumbatu, is also known for its skilled fishermen and many men spend years of their lives fishing at sea. My interviews suggest that most men in the region have worked as fishermen for at least some portion of their lives, and many men who have other primary occupations go on occasional fishing trips with a collective on a group-owned boat. Women do not generally fish at sea, but many collect shellfish and catch sardines in the shallow waters for home use and for sale; many women supplement their income by selling dried sardines, which are a flavorful addition to cassava boiled in coconut milk and to many other stews. When walking through villages, one often sees large mats strewn with the tiny fish drying in the sun. The region is also notable for its accomplished ship-builders, and many men make a living building dhows and smaller vessels.

Many people in the area live in modest homes of cement brick or coral brick; those with fewer resources build homes of wattle and daub. Cement and coral brick houses often have concrete floors and tin roofs, and the more modest homes normally have earthen floors and roofs thatched with woven palm fronds. The very poor may live in tiny huts made entirely of thatch. Nearly all but the smallest homes have courtyards at the rear with a kitchen and bathroom area off to the side. Washing and other chores are done in the courtyard, and the space often functions as the social center of the home. Building a house is considered a man's responsibility, and many young men postpone marriage until they are able to build a home suitable for a family. Most build with the help of friends and relatives, and many make the bricks themselves. A house may take several years to complete because building is normally postponed when funds dry up. Only a few homes in the area have electricity today, and slightly more have running water; those with running water normally have only one tap in the courtyard. Others take water from wells or public taps, and as elsewhere in East Africa, fetching water is considered women's work. Some houses, like the one I lived in, have built-in cisterns for water storage. Prior to my arrival in Kinanasi in 1999, there was no electricity in the village. When I arrived, I arranged with the owner of my rented house (who lived in town) to wire it for electricity instead of paying monthly rent.

Since the closest electric pole was about half a mile away, we also had to put in three new poles. Once the electricity was brought to that part of Kinanasi, the local mosque was wired, and a couple of my neighbors arranged for electricity in their homes. When I returned in 2002, the electricity was still running, but by 2005, only one or two houses still had electricity. The others had been shut off because no one paid the electric bill.

In 1999–2000 and 2002, I lived in the coral-brick house with a woman called Mwanahawa, who was introduced in chapter one. (In 2005, I stayed in the same house with another young woman, called Kijakazi.) When I was planning my move from town to Kinanasi in 1999, I told the regional administrator, who had been helping me find a house, that I was uncomfortable living on my own and would like to live with someone who could teach me how to cook local foods and help with housework. He had an ideal solution: his niece Mwanahawa, who was divorced and living in her mother's home in a village a mile away from Kinanasi, was available and would appreciate an income. The living arrangement worked out well. At about age 30, Mwanahawa was only slightly older than I, and she was intelligent, confident, and not at all shy about setting up house with a complete stranger. She moved in on my second day in the community, and quickly took control of the household. We had a similar sense of humor and enjoyed spending time together; Mwanahawa became a close friend and key consultant throughout my research and remains so today.

Although I thought we would share cooking duties, I found it difficult and time-consuming, and thus most of the cooking fell to Mwanahawa. I, however, did most of the shopping and other errands and paid her a monthly salary. She was an excellent cook and enjoyed preparing a wide variety of rich dishes; I loved her cooking, and gained about 10 pounds during the first few months we lived together. She found my subsequent attempts to lose weight through sit-ups and jump-roping very amusing, and insisted that I gained weight not from the food but because I drank too much water. Mwanahawa often joked that I was her "husband" since I went out to "work" every day while she stayed behind to take care of the house. Our relationship, however, was more like sisters. As the younger, I was expected to do errands and other tiresome tasks. As the elder, Mwanahawa was in charge of our household and took pride in running a very tight ship. She was well educated by local standards, a devout Muslim, and was considered by many people in the area to be admirably religiously learned. This was partly due to the fact that she was the youngest child of Shaykh Ahmad Mohamed, a man who was well known during his lifetime for his religious learning and teaching and was reputed to have died at age 120. When we first met, Mwanahawa had no children, and had been divorced by her first husband, whom she had married at a young age. She married again while we were living together, but her husband divorced her soon after and, as of 2009, she has not yet remarried.

Marriage is an important event for all women and men in Zanzibar, and few people never marry.[5] In 1999–2000, I interviewed approximately 75 women and about 25 men specifically about their marriages and divorces; I interviewed several more in 2002 and 2005. About half of the women were involved in marital disputes at the time of the interview, and the others had been divorced out of court at least once; most of the men had divorced a wife. As noted in chapter two, as a young, unmarried woman, it was difficult for me to talk to men about marriage and divorce, especially those my own age, and I did not discuss such topics with any but much older men. Consequently, I have fewer interviews with men about marriage, and most of these men were over age 50. The women I interviewed were between 18 and 90 years of age and had varying levels of secular and religious education. Although I expected a woman's age and religious education to influence her potential involvement in a dispute about an alleged repudiation, this was not the case; such disputes happened in all age groups, as illustrated by the age difference between Shindano, around 60, and Jabu and Zaynab, both in their early twenties.

Most people in this part of Unguja explain that a girl is marriageable once she begins menstruating and becomes a *mwari,* which might be translated as a physically mature, marriageable young woman.[6] At this time, a girl is no longer a child and is expected to behave with greater decorum and dignity. Ideally, she is supposed to cease attending wedding parties until she is married, and although she should keep herself immaculately clean and tidy, she should not beautify herself with henna or makeup (lipstick and black eyeliner are most commonly used) until the time of her own wedding.[7] However, in practice these ideals are often not observed, and one frequently sees unmarried teenage girls and young women at weddings (photo 3.1). Although girls are still considered marriageable once they become a *mwari,* today it is rare for a girl to be married right away. Both young women and men marry later than they did in the past, usually in their twenties, and it seems that this is at least in part a result of the great value placed on education and increased opportunities for pursuing education.[8] Most young people will not marry until they have finished or quit their schooling. In 2002, I asked

Photo 3.1 A Rural Wedding Celebration.

Bi Mboja, a neighbor in her forties who had relatively little education, what happens if a young man proposes to a girl who was still in school? She said,

> Now, it depends on the perseverance of that young man. If he already wants to marry her and she is still studying, then it is necessary for him to wait for her. But if he can't wait, then he will have to look for another woman and let the first one continue studying.

The court clerk Bi Hamida told me that she herself had refused several suitors by telling her parents that

> I had a goal to finish school, to get my letter to get work, and then I would get married. So when this one [her husband] came along, I didn't have any way to refuse—Almighty God had already written my fate, and I had work, I was already grown up at age 24, so why wouldn't I get married?

Most women over age 50 reported that they were married off as soon as they became a *mwari*, even if it was as young as 10 or 12. However, since few people, especially the elderly, bother to keep track of their exact age in years, it is hard to discern actual ages at marriage. Only one woman claimed she was prepubescent when she was married; however, she stayed with her parents until she became a *mwari,* and then went to live with her husband. Interviews suggest that being a *mwari* may have connoted more adult status in the past than it does now. One woman, Bi Mwajuma, said that although she married young when she moved to her husband's home she did all the cooking. I said "You did?" and she replied, "Well, of course! I was already grown! I was a *mwari!* I was grown up!" As an unmarried woman, I was often referred to as a *mwari.* This rarely seemed to be a simple statement of fact, but was rather used despairingly or critically. During my first trip, I was saved from too much pity by using my status as a student to explain why I had not married. I had less success explaining why I remained single on subsequent trips, and was often urged to marry.

Women of all ages reported that they had little say in their marriages, though I know of several young women who married men of their choice. Nearly all women over age 50 explained that they had no idea that they were getting married or to whom until the day of the wedding, and there is a fairly conventional way a woman describes her first marriage: she was out playing with her friends one day, when suddenly she was taken inside by adult women who began beautifying her with henna and then told her that her wedding was the next day. In Kiswahili, the verb *kuoa* (to marry) is used for men and the passive *kuolewa* (to be married) is used for women; thus a man "marries" and a woman "is married," and these are not interchanged. The verb *kuozeshwa* (to be married off) is the verb used to refer to parents marrying off children; during my research, it was common for women to use this verb even in marriages of their choosing, referencing the necessity of a guardian approving a woman's marriage. In describing the past, when it was common for boys to be married off by their parents, older men used *kuozeshwa* to describe being married off by parents in their youth. Some laughingly claimed to have been "seized" by parents who whisked them off to be married. Like women, many older men said they were married off at very young ages, and many women confirmed that they were married to boys not much older than themselves.

Although parents may still arrange marriages for children without consulting them, today it is far more common for a man to approach the parents with a proposal of marriage (*posa*). People meet or find a potential spouse in a variety of ways. Sometimes young people meet at school, or a young man may hear that a certain young woman is available and begin a pursuit based on her reputation. Other men may ask their mothers for suggestions of young women. Some young men stressed that the ideal spouse would be their father's brother's daughter; my friend Bilal and his male cousins often joked that they marry each other if one had been born female. However, this patrilateral parallel cousin marriage seems to be a matter of choice rather than necessity, and seems less prevalent than Middleton describes elsewhere on the Swahili coast (1992). Proposals may also come from a chance meeting or sighting. Bi Dana, a sharply intelligent woman in her late sixties, explained the betrothal process:

> Usually, what happens is that a young man…sees a girl who is a potential *mchumba* [fiancé, fiancée, potential marriage partner]. He says to himself, "That girl there, can she perhaps please me? Deep down in my spirit, I think I want her. I want to marry her so I'll propose." Then he would have to determine where her *kwao* (family home) was and he would have to find a local resident to take him to her; that is, if he didn't know where she lived. *Ehee!* (Right!) The locals would send him to her *wazee* (elders). Or perhaps the parents of the boy would go themselves. In that case, the

boy's father would say to her elders "He has seen that *mchumba* and he wants to marry her." Then the father of the girl would comment; if he agreed then they would be married off. If he didn't agree then the boy would find someone else.

Dana continued for some time, and explained that there may be difficulties if the girl lived far away because the young man's parents would not know her or her relatives. Therefore, it would be difficult to determine if the girl was a good person who was respected by others.

I also heard of some marriages that were the culmination of a long-blooming secret romance. Often, these stories involved the second or third wives of polygynous unions.[9] One of my neighbors, about 40, married his second wife in secret; she stayed at her home and he at his with his first wife. When the secret leaked out, it was said that the pair had loved each other since they were together in grade school. His first wife, failing to appreciate the romance, was furious. Middleton has also described these *ndoa ya siri* (secret marriages) among the people of Zanzibar and the Swahili coast (1992). In rural Unguja, although occasionally such marriages are regarded as charming romances, when they involve polygyny they are more often subject to criticism. People usually have a low opinion of a man who secretly marries a second or third wife.

When he wants to marry, a young man must get permission from his elders. Often, the permission of the paternal grandfather, as the head of the family, is sought before the father's permission. Because the young man's family often pays the bridal gift known as the *mahari*, young men are sometimes encouraged by their fathers to wait until the money is ready before making the proposal. Many men complained about the escalating amount of the average *mahari* and wondered how their sons would ever marry. After getting the approval of his elders, a man proceeds to the elders of the woman to propose; if it is not her first marriage, then the proposal may be made directly to her. If the woman's elders accept the proposal, then negotiations of the *mahari* and other elements of the marriage proceed. If it is a woman's first marriage, her grandfather, father, mother, uncle, or elder brothers will negotiate, but women who are marrying for the second or third time often negotiate for themselves.[10]

If a young woman's elders accept the suitor, she is generally expected to do so. I know of only very few women who went against their parents' wishes. Although women sometimes explained that according to *sheria za dini* they have the right of refusal, they always added that according to local norms, a woman must accept the suitor if her parents did so.[11] If it is not the first marriage, however, women can generally refuse suitors without much problem, even if it is contrary to the elders' wishes (see also Ingrams 1931, Caplan 1984). Among some women, however, the ideal of deferring to the authority of elders also extends to second marriages. Mwanahawa was one of these women. From the first day we met in 1999, she had a somewhat elusive fiancé who lived in Zanzibar Town. Earlier that year, he had approached her elder brothers and asked to marry her. They agreed, but the marriage took a frustratingly long time to come to fruition. We talked about it constantly. When she first told me about the proposal, she said that she did not think much of him, but would agree to marry him if her brothers wanted her to do so. She explained that since her father was dead, they were her proper elders and she would do what they wished even though she had been already married and could do as she pleased. I was perplexed, and asked why she would still accept their decision. She explained that since she was an *mtu wa dini* (a religious person) she must follow their requests; essentially, she held herself to a higher standard and believed a truly devout woman would accept her elders' authority in all her marriages. Several months and a visit or two from her fiancé later, I was certain that she liked him. However, she still claimed she was "unsure" about him and would only marry because her elder brothers thought she should.[12] They married in July 2000, but the union was short-lived; she was divorced less than a year later.

Divorce is very common in Zanzibar, and there is little stigma attached to it. Most divorced people remarry and I knew several women and men who had been married three, four, or even five times. Most divorces take place out of court through repudiation, often known as *talaka*, which is usually enacted through a spoken or written statement of divorce; spoken divorce is known as *kutamka talaka* and written is known as *kuandika talaka*.[13] In a divorce by repudiation, a spoken pronouncement is acceptable, but written divorces are preferable, even to the extent that an illiterate man might ask someone to write the statement for him by proxy. Shaykh Hamid once explained that although writing the divorce was not mandatory because anyone could destroy the paper, it was helpful to have a written document because otherwise witnesses were necessary. Mwalimu Adamu, a gregarious teacher in his forties, explained the benefits of writing repudiation similarly,

According to the essential nature of divorce, pronouncing the statement is sufficient. However, nowadays, in order to ensure complete proof and to prevent problems later on, it is better to bring witnesses to an actual writing of the paper; then there are no problems and you know that, truly, "I've divorced her."

Although most religious experts and scholars stress that a man should not repudiate his wife without just cause, they acknowledge that the matter is ultimately between an individual and God, and that they have no means of enforcing such a standard. In one of our many conversations about divorce, Shaykh Hamid once explained that, theoretically, men's right to repudiate their wives was rather limited.

Men will divorce their wives without any reason whatsoever. Now, truly, this is something that people do, but it is neither *sheria* nor *mila*. They just do it because they don't want her anymore. But it is not *haki* (i.e., one does not have the right) to divorce without a significant problem [in the marriage]. For example, if your wife left the home, you don't have the right to divorce her. You have to find her, then call together the *wazee* (elders) of both wife and husband, and ask why she left. You can't divorce her. They [the elders] tell her to return to you, and not to do it again.

In repudiation, a divorce is considered final if it has been delivered three times. After the first two statements, the divorce is revocable and the couple can resume living as husband and wife. In Zanzibar, it is fairly unusual and considered rash for a man to divorce his wives three times all at once. Most often, only one statement of divorce is written at a time, and is preferred as being the prudent, religiously appropriate means of enacting a unilateral divorce. In one of our interviews, the late Mzee Farad, a frail and reserved man of about 60, explained the importance of prudence in divorce.

If you divorce a woman three times, you can't come back to her. If you divorce her three times then right away it is all over. If you want to divorce her, divorce her only *once*. Then if your spirit moves you, you can return to her.... If you divorce three times at once, it is considered a prideful divorce. If you want to maintain civility, divorce only once. A prideful divorce is one of anger, and it is not good in Islam to divorce in anger all at once. It is better to do it one time and then if you want to, go back [to her]. Then, the second time, you can divorce her again. And then how many are left? One. If you divorce one more time after that, then it's all over—the husband can't return.... If he wants to return to her, he can't do so until she marries someone else and is divorced again. If she's divorced again you can marry her again, and you have to pay the *mahari* again.

As noted previously, it is not necessary for a woman to be present at the moment of divorce, and women may not know about the repudiation until they receive word from a third party. Considering this, it is perhaps unsurprising that many cases in the court centered on whether a divorce by repudiation took place.[14] When I asked people why these disputes occurred, many blamed men for trying to assert power over their wives or receive financial benefit. Shaykh Hamid and other *kadhi*s expressed their sympathy for women, and were quick to blame men for putting their wives in awkward positions by sending them away without writing or pronouncing a divorce.[15] Bwana Fumu, the chief clerk, thought that a man might send his wife away and cease to maintain her for financial reasons, but would not divorce her because he did not want to pay the balance of the *mahari,* which he would owe at the time of divorce if he did not pay it in full at the wedding.[16] This is not a problem that is limited to rural areas. In a 2005 interview at ZAFELA, I asked the young lawyers what they felt were the most pressing problems facing Zanzibari women. They answered that women's most serious difficulties involved divorce rights, and the first example they gave was that many women were divorced without their knowledge.

There are also other ways of divorcing out of court. Sometimes, a woman will secure a divorce through *khuluu* by compensating her husband and/or relinquishing certain financial rights, such as any unpaid *mahari.* This is known in the vernacular as *kununua talaka,* or "buying a divorce." Alternatively, a man might request money when his wife asks him to divorce her. Out-of-court divorces may also occur through a process known as *kuandikia pesa,* or "writing for money." This refers to a situation in which a man asks his wife for money when he divorces her *without* her request. It is thus similar to a man receiving money for divorcing his wife unilaterally through repudiation, and is not unique to rural Unguja. Stockreiter has found that the practice existed as early as the late nineteenth century in Zanzibar, as evidenced by *kadhi* court records from Zanzibar Town (2008: 230–231).[17] Caplan had described a similar practice on Mafia Island in which husbands asked their wives to "buy" their *talaka* (1995).

All lay persons I interviewed were familiar with the practice of "writing for money." Those who expressed an opinion about the practice were fairly critical of it, and some described writing for money as *mila mabaya* or a "bad custom." Others declined that the practice was *mila* at all, but was just bad behavior.[18] Tano, a woman in her early forties, was married and divorced three times and was very poor, even by rural standards. Tano's first and third husbands both wrote for money when they divorced her. The first husband asked for her *mahari* back, and she countered by saying that she did not ask him for a divorce so should not have to pay any money. She showed the paper he wrote her to her parents, who told her that she should not go back to him because every time she went, he tried to write her for money. Her parents told her that this was not lawful, and that it was simply his bad character that caused him to ask her for money. Tano told me that when her third husband wrote for money, however, she just paid him; she did not re-call the exact amount, but thought it between 5,000 and 10,000 shillings (8–12 USD). Other women also told me that they simply paid the money because their husbands asked for it.

Not surprisingly, no men claimed that they had "written for money," and many agreed that it was a blameworthy practice. When I asked the elderly Mzee Chumu, a spry and outgoing man with a large toothless grin, if he ever "wrote for money," he was aghast. In describing his divorce, he distanced himself from the practice by emphasizing that he had provided generously for his ex-wife.

> *Ataa!* [exclamation]! No, I didn't! Not one cent. Me, I took money to her [divorced wife] and sent her food for *edda* [the waiting period following divorce]—and even after she finished *edda*. We still got along, it was just that our time together was up—it was fate. Until now I still support her family. If her brothers come and ask me to help them I give them 500 or 600 shillings.

I also talked to several religious experts and teachers, *kadhi*s among them, about the practice. All considered it unlawful and shameful, and writing for money was never recognized as valid divorce by the court. There were varying ideas about why the practice occurs. Some explained that men "write for money" to facilitate a marriage to a new wife, which requires a new *mahari*. Another argued that the men and women involved in the practice do not understand the law. Yet an-other claimed that it is an issue of power and greed, and suggest that men understand the law but intimidate the women into paying money for male-initiated divorce. Shaykh Hamid explained the practice as a moneymaking venture. He referred to it as men "doing business," with women, and gave me an example of a man giving a woman 10,000 shillings for *mahari* but asking for 10 times that in divorce. He said, "If he demands that money from his wife, he will be abusing her." According to Shaykh Hamid, writing for money is not appropriate *khuluu* because an extrajudicial *khuluu* is only valid if a woman asks her husband for a divorce.

Lay people sometimes explain that a woman's payment of money to her husband is only necessary if she has been at fault in marital strife; in chapter six, we will see this understanding echoed in court. The problems between one young woman, Sikujua, and her husband illustrate this idea about payment. I met Sikujua, an attractive and amiable woman of about 25, when I was interviewing women in the *shehia* neighboring Mnazi Mrefu. Her maternal uncle, Mzee Bakari, heard that I was interested in marriage and divorce, and insisted that I talk to his niece, who was having marital problems and had requested a divorce because her husband Jafari was beating her. When we met, Sikujua was living with her mother. We talked on a couple of different occasions and Mzee Bakari, who was clearly very concerned for his niece and angry with her husband, talked with us. Sikujua told me that she had left her husband's home on what she claimed was both "her wish and his order." Jafari eventually came to her mother's home and asked her to come back. She told me that when she refused, he responded by saying, "Fine, so pay me." Sikujua explained that she would not pay him, specifically because he had sent her home to her mother, which was akin to divorce by repudiation. Mzee Bakari adamantly agreed, and claimed that Jafari had no right to ask for money because Sikujua was not at fault in the marriage. By abusing her and sending her away, it was Jafari who had erred. Sikujua would only have had financial responsibility if she had caused problems in the marriage.

Women's Experience of Divorce

In describing their experiences of divorce, most women explained that there was little communication and no mutual decision making in out-of-court divorces. Few women were informed in advance of their husbands' intention to divorce them, and none indicated that the matter was discussed beforehand. A common reply to my questions about why a marriage ended was "Well, the marriage was just over. *Basi*! (That's all!) There was nothing left to say to each other because the marriage was over." Or, "It was fate. Our love had ended." Bi Patima, in her seventies, said, "He just divorced me. There was no reason. He wrote the paper. No...I didn't feel bad. I just left. I took the children and went home. He just divorced me, that's all." Although some women like Patima noted that their husbands "wrote the divorce," others never heard the words of a divorce pronouncement or saw the written divorce paper. As a result, they did not get the kind of confirmation of divorce that is favored in court.

My interviews indicate that women often regard divorce as unavoidable. Many told me that they expected to be divorced at some point in their lives, and were well aware of the possibility that they would be divorced without their immediate knowledge. Even so, many women reported that they were surprised and upset when it actually happened. Bi Mwajuma, in her mid-forties, was divorced three times by her first husband. When I asked her if she questioned her husband about the divorce, she said, "He just divorced me without a reason and that was it. The way I see it, I was already divorced and I didn't have any reason to ask him—what would it change? I was angry, though, that he divorced me without a reason and never came back."

Bi Pili, a sprightly septuagenarian who had survived three divorces from three different husbands, told me that she was shocked when her second husband divorced her. She explained that she had gone with him to live on the island of Pemba, and the divorce happened when she returned home to Unguja to give birth.[19] After she had the baby, she waited for him to take them back to Pemba, but eventually she received word from a third party that he had divorced her. "I didn't know that I'd been given a divorce. I just waited for him. I had already given birth and he didn't come to get me to go back to Pemba." Bi Pili told me that she was distressed and wanted him back, but she never saw him again. When I asked her if she had ever considered going to find him, she said "No, if he has already divorced you then, *basi*, he won't support you again."

Other women of about Bi Pili's age suggested that later in life they learned to accept divorce as a matter of course and therefore learned to protect their material interests. Bi Tatu, a droll woman of 60 who had been divorced twice, explained that she was quite upset about the first divorce. She had married when she was very young, and was divorced by her husband after one year of marriage when she was seven months pregnant. She was shocked, and cried and refused to leave his house. Finally, her grandmother came to her and told her firmly that she must leave her husband's house because she was divorced. Tatu complained to her parents, but they counseled her to "forget it." They told her that her marriage was over because fate dictated it and there was nothing she could do. Her description of her second divorce 20 years later was quite different. She did not describe her emotional reaction, but rather her plans to ensure that her second husband did not take advantage of her financially.

Women and *kadhi*s like Shaykh Hamid tend to blame men for the prevalence of sudden divorces. They attribute it to the inherent irresponsibility of men and individual men's lack of commitment to support a wife. Most men, however, do not share this view, and tend to describe their use of divorce by repudiation as the reasonable outcome of problematic marriages. Take, for example, Mzee Omar, a man in his seventies who told me that he divorced one of his three wives because she was always away from home on errands or visits:

> If [I] went out for two or three days when [I] got back she wouldn't be there! She just went her own way and this did not make me happy...She'd just go everywhere. I'd get home and want food and she wouldn't be there. It was as if she wasn't my wife. She just went out to see her kids and wherever else.

He stressed, however, that he had divorced her properly: he wrote the statement of repudiation on a piece of paper and sent a child to deliver it to her. It is worth noting that his view of a proper divorce involved divorcing her from afar and sending the divorce paper by a third party.

Mzee Chumu told me that a man divorces his wife properly when he does the following: "He says his name, then 'I divorce you from today,' like the fifth month or the twelfth day, 'today I state that you are no longer my wife.'" When he divorced one of his wives, like Mzee Omar, Mzee Chumu wrote the statement out of his wife's presence and had a child deliver it to her. He said she cried when she got the paper and sent her older siblings to convince him to take her back, but he refused. When I asked how he knew that she cried, Chumu said that he had eventually gone to talk to her, "I went a few days later and I stayed for about four hours, and we talked about how we were leaving each other. She said she wanted me again and I told her 'it's over, we have to forgive each other now.'"

Like the wives of Omar and Chumu, many women do not witness the actual repudiation. However, all women experience similar structural events of divorce. The two most prominent are leaving her husband's home to return to her family, and the removal of her household goods, *vyombo,* from the husband's home. My research indicates that when these events occur, women often assume that a divorce has taken place even if they have no evidence of the act of repudiation.[20] For most women, the most definitive event of divorce is leaving her husband's home to return to her own family for shelter and support. Newly married couples in rural Unguja are often virilocal. Ideally, a man will build a new home for himself and his wife, but moving a wife into his parents' home is not unknown; some men, however, indicated that they would be uncomfortable with the latter. Because of the prevalence of virilocality, when a man tells his wife *nenda kwenu* (go home) she may interpret it as divorce, regardless of whether she receives a divorce paper or hears a pronouncement. When "going home," a woman usually goes to her mother and father, and if her parents are divorced, she will normally live with her mother or maternal kin. This is similar on Mafia Island (Caplan 1997) and in coastal regions of the mainland, where women maintain strong ties to their mothers' families that are an important source of economic and social support (Landberg 1986).

Recall that Jabu explained that she was "sent home" by her husband, and did not see him again for years. She believed that she was divorced (as did her parents and community), and remarried. Shindano also said that her husband told her explicitly to leave his home. Zaynab came to court and stated that she was divorced because her husband told her to leave his home and return to her parents. Several other cases in the Mkokotoni court were opened in the same way, and the focus on "returning home" is also common theme among women who never went to court. Bi Kombo, in her late forties, said that she was not told to go home, but was actually taken there by her husband. When her husband suggested they visit her parents, she was worried. He had not mentioned a divorce, but why else should he want to go to her parents? Kombo explained that when he took her home, he gave a paper to her father, who read it and confirmed her suspicion that she was divorced.

The significance of being "sent home" was prevalent even among adolescent girls. I came to know the girls of Kinansi quite well, and I noted that when talking about marriage and divorce, they stressed many of the issues that were important to the adult women. Our house was a popular gathering place for young people, and as a result I spent many afternoons and early evenings with them. At first, I attracted much interest as the only *mzungu* in the neighborhood, but Mwanahawa was also well liked by the teenage girls and they often came to see her. The large shady courtyard in the center of the house was an ideal place to spend time, and the space seemed to be always cool and breezy.

Our cistern meant we had a constant supply of water for anyone who wanted to wash clothes, and girls often came by to do their laundry. Our most constant visitor was rambunctious Rehema, who was about 15 when I first met her. She was short and stocky, quick to laugh and somewhat brash. I soon learned that she was the leader of the children: the boldest, the loudest, and the initiator of all games and fun. She was also the only girl I knew who could ride a bicycle, and she caused much excitement whenever she'd take off on mine with her ragged *kanga* flying. Mwanahawa and I grew very fond of her.

On some afternoons in the courtyard, conversation turned to marriage. One day, the girls and I talked about divorce, and they had much to say about what happens when a divorce occurs. Many girls focused on how quickly a woman could be divorced by her husband. Rehema exclaimed, "She could be divorced right there at her own wedding!" The girls explained that there is always much yelling and arguing in a divorce, and that eventually the woman goes back to her parents' home. Tana, a precocious, fast-talking girl of 11 with a flair for drama, eagerly acted out a typical divorce. She stood up, pretending to be a man, and yelled at her imaginary wife, *"Nenda kwako, na usije, usije, usije, tena!"* (Go home, and don't ever, ever, *ever* come back!). Her performance did not emphasize the written or spoken divorce statement, but rather the act of a man sending a woman back to her parents' home. This emphasizes the significance of the marital home as the territory of the husband, from which a woman can be driven. I recall that another little girl, Tulu, of about seven or eight years old, once told me that her mother had been "sent home" by her husband, but had recently returned. (Tulu also fancied that her tiny younger sister was "married" to another little boy, and that "they annoyed each other, because that's what married people do.")

One evening in 2002, at their request, I had been recording the kids reciting poetry and songs and playing the tapes back for their amusement. Rehema, then about 18, suggested acting out short skits (*michezo*) for my tape recorder, and they were very popular. After a few evenings, I had recorded about 20 short skits concerning family life. Girls took the lead in organizing the skits and play-acting, and the *michezo* took a raucous and comical look at marriage and family life. Several featured children asking their mothers for money, food, or clothing. Others featured husbands who complained that their wives were always roaming about, and wives complaining that their husbands stayed out all hours.[21] Most interesting were the *michezo* that mirrored the real-life situations that can result in a courtroom dispute over alleged divorce. In the first,

Rehema played a wife and her younger brother, Faridi, played her husband. He told his wife that he was tired of the constant presence of her girlfriends in the house. "Are your girlfriends *still* here?" he asked, "You don't listen! I'm going out on an errand and when I come back I'm going to chase you out and divorce you!"[22] In another, Rehema again played the wife, and a boy named Juma played her husband. The skit opened with Rehema sweeping and singing to herself. Juma knocked on the door, and greeted her, "Hello, my wife—" but before he could finish the sentence, Rehema cut him off: "Your wife?!" she yelled, "I'm not your wife! You divorced me two years ago. Get out! And, here, take your child with you!" After a pause, Juma told her he was leaving but that he was going to divorce her. Rehema answered, "*Bwana*, you've come back to me in a *bad* way!" He exited, and a girl playing a friend entered, stating that she just overheard that Rehema had been divorced. Rehema told her, "Well, yes, but that happened a long time ago!" The skit concluded with the friend commiserating that her own husband had thrown her out. In these skits, note that the children did not act out "writing" or "speaking" a divorce statement in any of them. Rather, they featured a man chasing his wife out of the home and sending her back to her parents.

A second important signifier of divorce that women emphasize is the removal of a wife's household goods from her husband's home. When a woman is married, it is considered her family's responsibility to furnish the house with supplies like pots, dishes, utensils, and washing tubs (see also Le Guennec-Coppens 1980). The word *vyombo* is used for goods of this sort and they are usually provided to a young woman by her parents and relatives, as Bi Mboja explained:

> In marriage, women must bring all the *vyombo*—all of the things to use in the kitchen. She is supposed to buy a bed; this is often bought for you at *kwenu* (your family home). Also dishes, an *mbuzi* (tool for grating coconut), mats for the floor; all of these things must be found for you at *kwenu*.

Women maintain ownership of the *vyombo* that they bring into the marriage and married couples in Zanzibar and elsewhere in coastal Tanzania do not normally hold joint property (Caplan 1984, Landberg 1986). I did not witness any marital dispute cases in which property was seriously in dispute. Because *vyombo* remain the wife's property, she takes them with her in divorce, as is also the case with Swahili women elsewhere (Landberg 1986). Thus, when a husband removes them from his house, or tells her to do so, women often understand this as indicating divorce. Several women told me that in a divorce a man will try to take everything from his wife—"even the clothes off her back!"—*except* the *vyombo*. Kombo, the woman whose husband suggested the "visit" to her parents, also highlighted the importance of the removal of the *vyombo* as an important sign of her impending divorce. She had suspected divorce even before her husband took her to her parents because she had been admitted to the hospital with a difficult pregnancy and when she returned home, all of her *vyombo* were out of the house. Kombo asked her husband where they were and accused him of theft. He denied the theft, and when she asked him again where the items were, he said, "Let's go to your parents." They went, and she was divorced.

In Shindano's description of the breakdown of her marriage, she repeatedly emphasized the importance of the removal of her *vyombo* from her husband's home as an indication that she was divorced. When questioning litigants about alleged divorces, the court staff regularly asked about the transfer of the woman's *vyombo* from one home to another. When I interviewed Shindano at her home about her case, the first thing she said was that her husband "cleared her out of" his home, and told her to take all of her *vyombo* with her: "That husband of mine, he told me 'get all of your *vyombo* out of here, and go home to your fellow dogs...!'" She also emphasized the importance of the *vyombo* when she talked about the response of community leaders to her plight. As is typical, she sought the help of the *sheha* before going to the *kadhi*. When she told the *sheha* and his aide that she had removed her *vyombo* from her husband's house, they said that they would try to prove that she had been forced to remove to *vyombo*. The *sheha* summoned Abu Bakr, and asked him to explain why he told her to remove her *vyombo*. He denied divorcing her and his response indicated the importance of the removal of the *vyombo* as a cultural signifier of divorce: he answered that he did not tell her to "clear them out entirely," but only to "set them outside." Therefore, he argued, this did not indicate any intent to divorce her.

The Case of the Foul-Mouth Husband:
Shindano and Abu Bakr

Let us now turn our attention back to the cases introduced earlier in this chapter. In all three, we met women who assumed they had been divorced when the structural events just described took place. One her first day in court, Shindano said that she was seeking a "registered" divorce, a court-issued certificate for a divorce that took place out of court. She claimed that her husband, Abu Bakr, had already divorced her, and that she had been having problems with him for quite some time, and he had not maintained her at all for the past several months. Shindano had already been to the *sheha* of her district to get help for the situation, but he had been unable to do anything for her. After explaining her situation to the other clerks, Bwana Fumu encouraged her to open a formal claim, and the clerks put together the *madai*. The document stated that she demanded a divorce receipt because she had been divorced by her husband when he told her, "Leave my house, and go home to your fellow dogs, slatterns, and lunatics!" Interestingly, it was also noted that in the eight months since her husband had sent her home, he had failed to maintain her; as we will see this detail became significant. The document concluded with a statement that the defendant divorced Shindano without giving her the proper documentation.

Madai

1. The plaintiff is a woman, aged 60, from K—.
2. The defendant is a man, aged 65, from K—.
3. They have been married for five years and have no children.
4. The plaintiff demands a registered divorce because she has already been divorced by her husband through being told "leave my house and go back to be with your fellow dogs, slatterns, and lunatics."[23]
5. The plaintiff claims that she was told these words about eight months ago; from the sixth month of 1999 until today, the defendant has not supported the plaintiff.
6. This suit was brought to court because the defendant divorced the plaintiff without writing her a paper or registering the divorce.
7. The claim originates from Northern A district, Unguja.
8. The plaintiff begs the court to order the following:
   a. The defendant must write her a paper of divorce immediately.
   b. The defendant must pay all court fees.

Despite her evident frustration, Shindano was a friendly woman. We talked about her case, and when we realized we lived near each other, she agreed to an interview the next week. I normally interviewed female litigants in court, but since we were neighbors, we agreed on an interview at her home. I arrived at her modest but immaculate house early the following Friday morning, accompanied by my research assistant, Ahmada. Shindano was pleased to see us, and greeted effusively with a bright smile. She had prepared a thermos of tea and purchased bread for our refreshment. We sat on a woven mat on the floor of the cool dark house, and Shindano immediately began telling us about her marital difficulties. While she was talking, a younger woman came into the house quietly and poured tea. Like many women, Shindano was not shy about discussing her marriage. She did not wait for my questions, but launched into a lengthy tale about her marital history and her husband Abu Bakr. She began her account much the same way as she did in the courtroom. She said she had been married, but had recently been "removed" from her husband's home when he told her to leave and take all of her *vyombo*. She told us about going to the *sheha,* and that they focused on the status of the *vyombo.* Throughout, Shindano expressed exasperation with Abu Bakr, confusion about her marital situation, and uncertainty about what would happen in court. Although she had asked for a divorce receipt, she suspected that she would not get one, indicating that she knew she had no proof of repudiation that would stand up in court. She felt a "heaviness of spirit" at the thought of being told to return to him, and hoped that he would pronounce a divorce in court. As with many women describing their marital troubles, she said repeatedly, "*Nimeshachoka*" (I'm tired, already)!

I asked Shindano if Abu Bakr was her first husband, and she told us that she had been married twice before. She had two children from the first marriage, but was divorced when they were very small. Now, both the children were grown and married, and she told us that the young woman who served us tea was her daughter-in-law. When I asked her about the first divorce, she explained that her husband simply tired of her, and sent her home with a divorce paper. She remarried soon after her baby son was weaned. She said that the second husband also divorced her, but as with Abu Bakr, much confusion surrounded it. She had gone home to her parents for a short time, and when she prepared to go back to her husband, someone came to tell her not to go back. The messenger did not give details, but simply said that there were some problems there. A few days later, the same messenger told her again not to go back, and she never did.

When she described her marriage to Abu Bakr, she focused only on the problems, and had nothing positive to say. She said she had not known him well when they married, and once she moved into his house, she realized she had made a mistake. When her new neighbors told her, "Oh, you really got yourself into it," she asked why, and they told her that he had many problems: he was foolish, had a tendency to gab and scold, and was generally unpleasant. She had problems with him from the day they married, and regretted it. When I asked her if Abu Bakr knew she had gone to court, she answered smugly that he would only know when he got the summons.

Abu Bakr was indeed summoned, and a week later he came to court. I recognized him as one of my neighbors, and although I did not know him well, he greeted me pleasantly when he arrived. A clerk read Shindano's *madai* aloud, and asked for his response. He said simply, "It's not true." He paused, and then carefully explained the situation from his point of view. He said that he had neither divorced Shindano nor verbally abused her, and suggested that she had fabricated the story in order to get a divorce from the court. He admitted that he had not supported her for the last eight months, but argued that this was because she had forbidden him from coming to her house. His *majibu* asked the court to rule that they were still husband and wife.

Majibu

1. The defendant agrees with points 1 and 2 of the *madai*.
2. The defendant does not agree with point 3, and he explains that he does not think that they have been married any less than 11 years.
3. The defendant does not agree with point 4 of the *madai*, and he explains that he has not divorced his wife as she claimed. He denies that he told her the words presented in her claim, and says, "I would never be able to say such words to her, not even [for] one day." The defendant claims she fabricated the story in order to get a divorce.
4. The defendant agrees with point 5 of the *madai* and explains that defendant has no problem with his wife but he was not able to go to her because the plaintiff told her not to go to her home to see her. She said this in front of other people when the defendant went to see her to try to resolve the situation.
5. The defendant does not agree with point 6, and says that he did not divorce his wife through either writing or speaking a pronouncement or saying anything that she might interpret as a divorce.

6. This claim originates from K——, Zanzibar.
7. The defendant begs the court to rule the following:
   a. The court throws out the plaintiff's claims because the defendant has no problem with her and she is indeed still his wife.
   b. The plaintiff must pay all court fees.

A few days later, they were in court together to see the *kadhi*. In her testimony, Shindano emphasized that they had had many problems in the past, but it was "different this time" because Abu Bakr had specifically told her to leave his home. When Shaykh Hamid asked if she had written proof of the divorce, she said she did not, but emphasized that her husband sent her away from his home and told her to remove all of her *vyombo*. When it was Abu Bakr's turn, he testified that he had not divorced her and had never spoken the insulting words noted in her claim. Furthermore, he was prevented from maintaining her because she had told him to stay away.

After hearing their testimonies, the *kadhi* explained that the plaintiff and defendant were not in contradiction. He said that this was because Shindano did not claim that she had heard or seen a divorce statement, and Abu Bakr confirmed that he had not made one. Therefore, Abu Bakr had not divorced Shindano through repudiation. Sometime later, when I asked Shaykh Hamid why he had not called any witnesses, he answered simply that he had not done so because both parties "agreed." He noted my puzzled expression, and explained patiently that when Shindano gave her testimony, Abu Bakr essentially agreed with what she said, so it was not necessary to call witnesses. In this case, he assured me, both parties were telling the truth.

Although the *kadhi* had determined no divorce took place, he did not dismiss the case, but rather changed the focus to the marital problems that he believed were at the heart of the dispute. Recall that although Shindano's *madai* was written as a request for a divorce receipt, the clerks had included in her claim the fact that Abu Bakr had not supported her for several months. Considering this and Shindano's testimony, the *kadhi* reframed her complaint as essentially a request for her right to maintenance. Shindano did not contest this focus on maintenance, but instead moved with him to a discourse of marital rights by reemphasizing her husband's failure to maintain her properly. In his ruling, Shaykh Hamid did not mention the alleged divorce at all, but rather gave the couple terms of reconciliation, *masharti*. He instructed Shindano to return to her husband, to cooperate with him, and to inform him when she was leaving the home. Abu Bakr was ordered to support his wife "according to the law" and to cease "cursing her with foul language." As usual, Shaykh Hamid wrote a provision for divorce if either party broke the terms.

Hukumu

Terms for the Plaintiff

1. Listen to your husband and live well with him in the house that he provides for you; do not run away from him any more.
2. And when you go to the farm, you must return to the house, and don't wait until nightfall to return.
3. And if your husband gives you food, cook it for him; also speak well with him and do not be harsh.
4. Consult your husband's opinion in everything you do, and tell him when you leave the house.

Terms for the Defendant

1. You must support your wife lawfully; give her food, clothing, and money to buy soap and other things.
2. Speak well with her, and do not be harsh, do not call her cruel names any longer, nor should you curse her, or use bitterness with your wife.
3. Live well with her, both at night and during the day, and fulfill all of these terms and if you do not, your wife will be able to end the marriage and she will be divorced.
4. And if either party breaks the terms, then they must report it to the *sheha*, and the *sheha* will write a report and bring it to the court; if the person at fault is the plaintiff, then she must buy her divorce as according to the law.

Bwana Fumu read the ruling aloud, and Shaykh Hamid explained to the litigants that Shindano had come to court to claim her rights from her husband, and emphasized again that they were not divorced. Although Shindano looked displeased, she did not protest and left the court with Abu Bakr. I did not see them in court again. As was our usual practice, Shaykh Hamid and I discussed the case after everybody left. In his view, he said, the case was not really about divorce. Rather, Shindano had come to court to complain about maintenance and verbal abuse.

"Go Back to Your Fellow Dogs": Zaynab and Rashidi

Zaynab's case was similar to Shindano's in many ways. When she came to court the first time, Zaynab and her father (who always accompanied her) explained that she was divorced because her husband, Rashidi, had ordered her from his home and told her to "return to her fellow dogs." Like Shindano, Zaynab had no proof of repudiation, and the clerks told her they should open a case and summon Rashidi. Zaynab and her father agreed, and the clerks prepared her *madai,* which stated that she was seeking a divorce receipt because Rashidi had divorced her. The claim, like Shindano's, also stated that he refused to support her after sending her home, that he refused to give her a written divorce paper, and cited his alleged abusive language. In this case, however, Bwana Fumu worded her demand as "one divorce" from her husband instead of a divorce paper, which might have reflected his doubt about the alleged divorce (see 7a).

Madai

1. The plaintiff is a woman aged 20, Mtumbatu, from M—.
2. The defendant is a man aged 35, Mtumbatu, from M—.
3. That the plaintiff and defendant are wife and husband and have been married for four years and they have two children; one died and one is still alive. That the plaintiff demands a [registered] divorce because she claims she is already divorced by her husband because she was told "Leave my home and go back to yours with your fellow dogs." It happened in the eighth month at 8 o'clock at night in this year, 1999.
4. That the plaintiff claims that for this entire period her husband has not supported her or their child.
5. The basis of this claim is that the husband did not give the plaintiff the divorce paper after he spoke the words that are cited above.
6. This claim originates from K—, Zanzibar.
7. The plaintiff begs the court to rule that:
    a. The defendant divorce his wife one time.
    b. The defendant pays court fees.
    c. Any other orders issued in the agreement of the plaintiff.

Zaynab was a lean, excitable young woman who was not at all shy about speaking her mind. On the many days they spent in court, she often became very agitated and spoke at a rapid clip. Her father was of a similar temperament, and it was clear as the case progressed that the clerks found the pair irritating. Even the mild-mannered Shaykh Hamid expressed impatience with them a few times. When I interviewed Zaynab, I learned that she was Rashidi's second wife. They lived in a large village spread out on the sunny east coast of Unguja, where Rashidi worked as a fisherman off the blindingly white beaches. Although only a few miles away, this part of Unguja is very different from the area surrounding Mkokotoni. It receives less rain, and farming is more difficult. One sees few large trees other than coconut palms, and other crops are sparse.[24] Zaynab explained that Rashidi had recently ordered her to leave his home and return to her parents, and had done so using foul language. After he sent her away, she returned to his house to retrieve her *vyombo*. While there, she asked him for a written statement of divorce, but he refused. She said that he had not maintained her or their child at all since she was sent home; as in Shindano's claim, this was cited in her *madai* as further evidence of her divorce. It was clear that to Zaynab and her father these events indicated that she had been divorced.

Rashidi came to court a few days later. He was tall, with a quiet but steely manner. When the clerks read Zaynab's *madai*, he claimed that he had not divorced Zaynab and argued that most of what she said was untrue. He explained that she went home on her own initiative: he had not sent her home and had "not pronounced anything," which meant that he had not divorced her. Therefore, he concluded, he could not possibly give her a divorce paper because she was not divorced.

Majibu ya Madai

1. That defendant agrees with points 1–3 of the *madai*.
2. He does not agree with the explanation presented in point 4.
3. He contests point 5 and answers that he has not told the plaintiff to "return to her fellow dogs"; and has not pronounced anything; the plaintiff went to her home of her own accord.
4. Concerning point 6, the defendant answers that he is not able to divorce his wife because he has not pronounced anything, as is stated above.
5. Concerning point 7a and b, the defendant does not agree with her requests because of what is stated above.
6. The defendant begs the court to issue a judgment for the plaintiff to follow what is stated below:
   a. The plaintiff returns to her husband's home immediately.

  b. The plaintiff pays court fees herself.
  c. Any other orders issued in agreement with the defendant.

When the two came to court together, Shaykh Hamid listened to their accounts of the problems with the marriage. After several very long and tiring days in court during which the litigants reiterated their initial claims at length, the *kadhi* explained the essential problem: Zaynab claimed she was divorced, and Rashidi said she was not. To the *kadhi,* this was different from Shindano's case because although Zaynab continued to insist that repudiation had occurred, Shindano had not. As a result of their disagreement, Shaykh Hamid called witnesses. Rashidi's witness was an elderly relative who said simply that Zaynab had not returned to her husband's home. At this, Zaynab cried out that she did not want to return to him and wanted her divorce paper. Her outburst prompted Shaykh Hamid to explain, as he put it, the applicable *sheria*: if she had not been divorced and refused to go back to her husband, then she would have to buy her divorce in *khuluu*. Zaynab countered by emphasizing Rashidi's verbal abuse, and Shaykh Hamid told her that if that was her major complaint, then she must go to the police to resolve the matter. He said she had two possible *sheria* (here, meaning "legal paths") if it was established that she was not divorced: she could return to Rashidi or she could buy a divorce from him, "as explained in *surat al-nisa* in the Qur'an."[25]

To no one's surprise, Zaynab's witness was her father. He testified that she had never received her divorce paper from her husband, and reemphasized that she could not go back to her husband because he had called her a dog. He added that in the past Rashidi had "written her for money" in the amount of 80,000 shillings, but they had not paid him. Rashidi interjected to explain that he simply wrote a letter telling Zaynab that he would divorce her if he received money, but he had not done so because he he they did not give him any.

Because there were no witnesses to a divorce or other proof of repudiation, Shaykh Hamid told the litigants that there was no way to prove that it had happened. As in Shindano's case, he proceeded to handle the dispute as a wife's claim for maintenance, and moved from talk of divorce to a discourse of rights: he outlined each party's role in the marriage, and instructed Rashidi to properly care for his wife. The *kadhi* told Zaynab that she must return to her husband or buy her divorce. Zaynab and her father were not at all happy, and she continued to press angrily for a court-ordered divorce. Like Shindano, she utilized the discussion of marital rights and duties that Shaykh Hamid had begun, and quickly reframed her argument to reflect this. She dropped her emphasis on the structural event of being sent home, and instead argued that the severity of the verbal abuse she suffered violated her wifely rights and justified a court-ordered "free" divorce (*fasikhi*). Shaykh Hamid calmly replied that although she did indeed have the right to sue her husband for calling her a dog, she must do this at the police station because any form of abuse was a criminal matter and was hence under the jurisdiction of the police and the secular court, not the Islamic court. Zaynab was not interested in taking the claim to the police, however, and reiterated her demand of a divorce. The *kadhi* was irritated, but told her once again that if abuse was her primary complaint, she must go to the police.

Shaykh Hamid wrote the ruling as terms of reconciliation that were similar to Shindano's. He required Zaynab to return to her husband, and required Rashidi to support her according to "laws of marriage"; the ruling also stated that if either party broke the terms, they would be divorced.

Hukumu

After listening to the claims of the plaintiff and the defendant, together with their witnesses, the court has decided that both the plaintiff and the defendant have made mistakes and the court issues the following terms.

Terms for the Plaintiff
1. The plaintiff is ordered to return to her husband's home.
2. The plaintiff must fulfill the laws of marriage to the defendant who is her husband by marriage.
3. The plaintiff must listen to and serve her husband when he asks her to do so.
4. The plaintiff must cease using foul language with her husband.
5. The plaintiff must speak nicely with her husband.
6. The plaintiff must fulfill these terms and if she does not she must buy her divorce.

Terms for the Defendant
1. The defendant is ordered to take his wife into his home.
2. The defendant must support his wife with enough food every day.
3. The defendant must give his wife two sets of clothing, two dresses, two headscarves, and shoes to visit people with. The defendant

must also give her money for incidentals and to buy items for personal care and beauty like perfume, powder, and bangles. Also, he must not use foul language. You are expected to fulfill these orders and if you do not your wife will be able to dissolve her marriage in a *fasikhi* divorce. The plaintiff and defendant are expected to send any further problems to the *sheha*.

Perhaps unsurprisingly given Zaynab's passion and stamina, this was not the end of the matter. The party was back in court a short time later. Zaynab had not returned to her husband, which broke the terms of the ruling and required her to "buy" her divorce. The *kadhi* noted this development in the case file:

In the court of the *kadhi* of Mkokotoni, the plaintiff, the defendant, and their elders are here and the *kadhi* has read them the decision of this case concerning their children. Although they were given *masharti,* these young people did not fulfill the terms after being explained by the court that the one who breaks the terms must fulfill the decision of the court and buy her divorce. The plaintiff has agreed to buy her divorce.

After a few weeks, Zaynab brought 45,000 shillings to pay for the divorce, and the matter was finally settled.[26]

The plaintiff and the defendant are here in court. After the plaintiff failed to fulfill the terms that she was given by the court, she must buy her divorce with the money for which she was married. And today in court, she gave this money to her husband in order to get her divorce. This is 45,000 shillings, which she has given him in order to get her divorce. The defendant divorced his wife after receiving this money. This case is closed with his receipt of the money.

Jabu's Two Husbands: Jabu and Rajabu

Although Jabu's case also involved a disputed divorce, it differed significantly from those of Shindano and Zaynab because she had remarried following the alleged divorce. Jabu, a poised and elegant young woman who seemed more mature than her 25 years, was from the same village as Zaynab on the east coast of the island. She came to court because of a dispute with Rajabu, her first husband. Jabu said she had been divorced three years earlier when Rajabu sent her home to her parents and ceased maintaining her. Some time later, she married another man. After hearing this news, Rajabu came to her claiming that he never divorced her and demanded that she return to him. When she refused, he went to the Chief *Kadhi* to request an order stating that she was not divorced and must return to him. In a procedural anomaly, the Chief *Kadhi* heard his plea and gave him the order requiring Jabu to return to him. She was forcibly returned to him, but she ran away and came to the Mkokotoni court to plead her case and request a divorce receipt.[27]

Like Shindano and Zaynab, Jabu's argument focused on the experiences she interpreted as divorce: he had sent her home to her parents, and had subsequently ceased to maintain her. The clerks were puzzled with her claim, however, and consulted with Shaykh Hamid before opening a case. The *kadhi* explained that could not reopen a matter on which the Chief *Kadhi* had already ruled, and told Jabu she must sue for divorce from Rajabu if she wanted to open a case in his court. She did so, and on the advice of the court clerks, her *madai* was prepared as a simple request for divorce on grounds of incompatibility. The *madai* did not mention that she had been divorced or that she had married another man.

Madai

1. The plaintiff is a woman, aged 25, from M——.
2. The defendant is a man, aged 45, from M——.
3. The plaintiff and defendant have been married for the past 13 years and have 3 children.
4. The plaintiff demands a divorce from her husband because she does not want him and no longer wants to live with him as his wife.
5. The basis of this claim is that the plaintiff does not want the defendant her husband.
6. The claim originates from M——, Unguja.
7. The plaintiff begs the court to rule that:
   a. The defendant divorce his wife and give her the divorce receipt because she does not want him any longer.
   b. The defendant must pay the court fees.
   c. The defendant must obey any other orders of the court.

Rajabu was summoned and came in a few days later. He was quite a bit older than Jabu, and seemed confident and worldly but not unpleasant.

He was a fisherman, and clearly a successful one because he came to court on a scooter. Very few people in the area had any transport other than the old-fashioned but sturdy Chinese-made bicycles. The clerks read the claim, and then prepared his *majibu*. Rajabu seemed convinced that Jabu had married her new husband unlawfully, and although it was not mentioned in the *madai*, his counterclaim noted Jabu's second marriage many times: Rajabu could not agree to divorce his wife because "she had unlawfully married another man," and had thus "broken the laws of Islam." He asked the court to order her to return to him as her "true" husband.

Majibu

1. The defendant agrees with points 1–3 of the *madai*.
2. The defendant does not agree with point 4 of the *madai*, and he replies that he will not divorce his wife because she married another man before being divorced by the defendant.
3. The defendant answers point 5 of the *madai* by stating that the plaintiff is demanding a divorce because she has behaved shamefully and broken the laws of Islam by marrying another man before being divorced from her first husband.
4. The defendant says that point 6 is true, about the origin of the claim.
5. The defendant does not agree with point 7 and with what the plaintiff demands from the court, and indeed, it is she the plaintiff who broke the law by marrying another man before her divorce.
6. The defendant begs the court to rule the following:
   a. The court disregards the plaintiff's *madai* because she has no basis for such claims.
   b. The court orders the plaintiff to return to the defendant if it is established that she is still the wife of the defendant.
   c. The plaintiff must pay the court fees.
   d. The plaintiff must obey any other orders of the court.

In their hearing before the *kadhi*, Jabu began by repeating what the clerks had set out in her *madai* and said that she was suing for divorce based on incompatibility with her husband. Then, however, she explained that the clerks told her to do this, and that in truth, other matters were at stake: namely, that Rajabu had divorced her years ago when he sent her home to her parents. In his testimony, Rajabu countered that they simply "left each other" when she went back to her parents. He said that she had once asked him for a divorce, but he did not give it to her. At this, Jabu asked him why he was lying, and Rajabu contradicted himself by replying, "You should have asked me for a divorce before going off and marrying another man!"

Shaykh Hamid listened sympathetically to Jabu's account of the alleged divorce, but continued to proceed with the case as a suit for divorce. At his request, the litigants brought their elders to court. However, he did not ask them to testify as witnesses. Rather, he explained that Jabu and Rajabu had a simple problem that required a simple solution: Jabu did not want to return to Rajabu as his wife, but Rajabu did not want to divorce her. Therefore, the *kadhi* decided that it was appropriate for Jabu to buy her divorce. Everyone agreed that this was an acceptable solution, and the elders began a lengthy deliberation over the appropriate amount of compensation. A mild argument ensued over what Rajabu's elders wanted and how much Jabu's family could afford to pay, and the situation became increasingly tense. However, in a move that surprised everyone, Rajabu suddenly announced cheerfully that he did not actually want money and would divorce Jabu for "free." It was a dramatic moment, and everyone, including the *kadhi* and clerks, were pleased with what they regarded as a gracious gesture. No one was happier than Jabu, however, and she and Rajabu parted amicably. This was a settlement rather than a ruling, and hence Shaykh Hamid did not prepare a *hukumu*; the clerks simply noted that Rajabu agreed to divorce Jabu in the case file. I never learned what sparked Rajabu's change of heart, since he left the courtroom quickly and I did not see him again.

Although Jabu followed the advice of the court clerks and opened her case as suit for divorce on grounds of incompatibility, her narrative of the alleged divorce still played a part in the proceedings. The *kadhi* could not uphold Jabu's allegation of divorce without proof, but he did not discredit her experience of the events that led her to believe she was divorce. Although he normally encouraged reconciliation— even when couples had been living apart for a long time—in this case he deemed it impossible because Jabu had married another man. He could not validate the second marriage because he had no proof that Rajabu had divorced her, so Shaykh Hamid decided she should buy the divorce. Moreover, he disregarded Rajabu's repeated assertions that Jabu had unlawfully married her second husband and therefore committed the criminal offense of adultery. He did not encourage Rajabu to pursue the matter with the police, as he did in other cases when litigants brought forth potentially criminal matters, like when Zaynab cited verbal abuse. According to Shaykh Hamid, there was no unlawful intent in Jabu's second marriage because she believed she had been divorced, which nullified the potential criminality of the act.

Conclusion

The many disputes concerning alleged out-of-court divorces appear to stem from the different divorce experiences of women and men, not from different understandings of what constitutes a legal divorce. Men in Zanzibar have far more power in enacting divorce than women because of their right to unilateral repudiation, and many such divorces take place out of the wife's presence, as illustrated in interviews with men and women. As a result, women interpret the structural events associated with divorce, leaving the husbands' home and/or removing her *vyombo*, as indication that repudiation has taken place. Shindano, Jabu, and Zaynab emphasized one or both experiences as evidence of divorce when narrating their experiences in court. However, women do not usually claim that these events are the equivalent of Islamic repudiation. Most women, even those with little education, understand what a lawful Islamic divorce through repudiation entails: a written or spoken statement of divorce. Bi Mboja once told me that the prevalence of disputes about divorce was due not to misunderstandings of Islamic laws of divorce, but rather to lack of communication between spouses; "There is no stranger to marriage here," she said, "not even one!" Even among lay people, marriage is often discussed as a body of rights and duties, and Mboja's comment meant that all people understand the laws of marriage and divorce.

By seeking a divorce receipt in court, women seek to legitimize the structural events of divorce. The prevalence of these disputes does not indicate that Zanzibari women are not familiar with Islamic divorce. Legal and religious knowledge concerning marriage and divorce is not gendered in the sense that men and women have different access to or interpretations of such knowledge. Rather, a woman's request for a divorce receipt shows how she views an unregistered divorce: the structural events of divorce are there, but the "proof" of repudiation is not.[28] However, as we have seen, the structural events of divorce alone are not upheld by the *kadhi* as legitimate divorces. Shaykh Hamid's ultimate goal is to preserve the marriage bond, and in these cases he makes a clear distinction between what is properly Islamic and what is not. When Zanzibari women explain their problems in court, they emphasize events that signify divorce. When she seeks a divorce receipt based on the occurrence of the structural events of divorce, a woman is referencing not only Zanzibari norms of marriage and divorce, but also Islamic law concerning proper repudiation and state law concerning the registration of divorces. She knows that the structural events of divorce are not an actual Islamic divorce, but a sign of divorce. Moreover, when women do not have evidence of repudiation for the *kadhi,* they may reframe the core issue of the dispute and move with Shaykh Hamid to a discourse of legal rights and responsibilities in order to claim maintenance or request a court-ordered divorce.

In interviews, many indicated that women are more able and more likely to go to the *kadhi* today than in the past. Many women regard increased access to courts in a positive light as a way to assert more control over the marital relationship. Others, however, expressed reluctance to go to court and, of course, most women who experience marital difficulties never go to court. My housemate Mwanahawa experienced a bitter divorce not long after she married in 2000. She was pregnant when her husband divorced her after less than a year of marriage, and she delivered a health baby boy not long after. As of 2008, the child's father had visited only once, and had not paid for any of his support, which he is lawfully required to do. Although Mwanahawa knows she can sue for child support in the *kadhi's* court, she remains reluctant to do so. She explained that she did not want to air the problem in public, although certainly everyone in the area knew her former husband was not supporting the child, and most people had a negative opinion of him. When I asked her if she wasn't angry and frustrated with him, Mwanahawa agreed that she was, but that "God would take care of the matter." Annelies Moors has noted that Palestinian women sometimes relinquished similar rights in order to enhance their moral standing in the family (1995, 1999).

Some scholars have sought to correct stereotypes about biases in Islamic law by noting that women tend to win most of their cases in many Islamic courts (Hirsch 1998, Wurth 1995). It would be difficult to say that these three women "won" their cases. With Shindano and Zaynab, the *kadhi* sought to address the problems he considered to be at the heart of the marital trouble and to remedy the situation through reconciliation. Jabu's case was different because she had married another man. Although Shaykh Hamid did not validate Jabu's alleged divorce, he did rule for immediate *khuluu*. In addition, he did not consider the second marriage adultery because Jabu thought she had been divorced. Therefore, women do not "win" these cases by getting the divorce certificate they originally sought: the *kadhi* would not uphold an alleged divorce without proof of lawful Islamic repudiation, and the women clearly did not think that the structural events they experienced were the equivalent of a lawful divorce. However, taking the matter to court gives women a forum to highlight the husband's neglect of his martial duties. As Erin Moore has demonstrated in her work in India, women's legal action challenges the boundaries of male authority both in and outside of the home—even when women are not fully successful in their claims (1994). As in the examples presented in this chapter, women use the courts to make lawful divorce-related actions that occur outside of the court. Although they may not always get what they initially sought (a divorce receipt in these cases), by taking the matters to court and working with the *kadhi,* women hold men accountable for failings in their marital duties.

CHAPTER FOUR

A Wily Wife and a Headstrong Husband: Determining Intention

In this chapter, we continue to look at disputes about alleged out-of-court divorces. Now, however, we turn our attention to cases in which the dispute centered not on whether a divorce occurred, but whether the divorce-action was valid. In Shaykh Hamid's court, establishing the validity of divorce-related action was not simply a matter of determining whether it had taken place, but hinged on whether the proper intention was behind it. Intention, or *nia* in Kiswahili (from the Arabic *niyya)*, is an important concept in Islamic thought that has great practical and legal relevance, and establishing the presence or absence of intention in legal acts was a critical element of judicial reasoning for Shaykh Hamid. In the cases examined in this chapter, we will see that the *kadhi* assessed the intention of the actors involved by considering the range of possible meanings of particular actions. To assess the relationship between outward actions and elusive inner states, he considered the various scenarios in which a divorce-related legal action can occur, and took into account changing gender roles in marriage and individual agency in determining marital status. Brinkley Messick has explored the divergence of opinion on the existence of intention in different kinds of legal acts in the writings of jurists of the Yemeni Zaydi school of Islamic legal thought (2001). He observes that "Given the assumed gap between forms of expression and intention, legal analyses amount to attempts to erect bridges from the accessible to the inaccessible. The interpretive work of evaluating spoken and written expression...represents such a bridging effort" (178). In the remainder of this chapter, I aim to explain one *kadhi's* practice of bridging.

The necessity of proper intention is emphasized in the performance of all Islamic ritual act and duties. It is also an important legal principle that is applicable to contract law, criminal law, and family law. Although opinions differ among legal scholars and schools on the importance of attention in various dimensions of law,[1] my concern is how the concept is used in everyday life and legal practice. In Zanzibar, lay people and legal professionals, men and women alike, agree on the importance of intention in religious and legal actions. People often emphasize their intent to perform particular acts, religious or nonreligious, in the future. For example, in a conversation about her religious practice, one elderly woman told me that although she was not presently praying on a daily basis, she certainly had the *intention* to start doing so, thus indicating that even though she was not praying she was not blatantly shirking religious duty since she might start doing so any day. Many individuals talk about religious study the same way—it was not that he or she did not want to study, rather the intention was there but the opportunity had not yet arisen. Similarly, references to intention sometimes pepper conversations about divorce, particularly among men. When I asked about personal history of divorce, men and women followed linguistic conventions of Zanzibari Kiswahili by infrequently answering the question with a simple "no." Men often used the word *nia*, by including a phrase such as "*sijakuwa na nia kumwacha* (I have not [yet] had the intention to divorce her). A woman might answer with "my husband has not yet divorced me" or "I have not yet asked my husband for a divorce."[2]

In a number of disputes, Shaykh Hamid placed great emphasis on determining if an alleged divorce was valid, and assessing the validity of claims of divorce hinged on determining the intention of the actors involved. If a man utters a divorce statement without intention does it count as valid divorce? If not, how can the presence or absence of intention be established? Shaykh Hamid's determination to assess the legality of out-of-court divorce was demonstrated by the way he brought the question of validity to the fore even in cases in which the litigants themselves did not dispute the divorce action. Consider again the example of Machano and Aisha from chapter two. Machano claimed that Aisha had remarried after he divorced her but before her waiting period, *edda,* was completed. In this case, Shaykh Hamid gave greater emphasis to the question of whether Machano's divorce action was valid than to the issue of whether Aisha had properly waited out *edda* before marrying another man. This was to ensure that Machano had not divorced her through the unlawful practice of "writing for money." In other cases, it was clear that Shaykh Hamid focused on validity because he assumed that some people had little religious learning and did not properly understand how a divorce must be enacted. Shaykh Hamid was not the only *kadhi* who held this sentiment. Another with whom I worked, Shaykh Vuai, also stressed establishing validity in the divorce-related actions. Like Shaykh Hamid, he said that this was because some people did not understand how to divorce, but also because he suspected that men frequently lied about divorcing their wives.

Anthropologists studying law and Islam have been concerned with cross-cultural variation in ideas about intentionality, the way in which individual motivations and inner states are assessed, and how intentionality plays out in legal contexts and dispute resolution. Lawrence Rosen, who has been most prolific on the subject, has considered the way people in Morocco asses the inner states of others, and suggests that in the Moroccan context the link between action and intention is thought to be evident on many levels, and an important means of assessing intent is through understanding "situated acts—occurrences that draw together the qualities of nature, background and biography to make an inner state 'obvious'" (1989: 53). For example, Moroccan judges told Rosen they could discern intention from inquiring about an individual's "background, relationships, and prior behavior" (1984: 53). This seems similar to the way in which Shaykh Hamid considers the range of meanings of particular actions based on social circumstance. As in the Moroccan context, Shaykh Hamid presumed that an individual's intention can be discerned. The *kadhi* recognized that actions are often ambiguous, and what a person does and what a person means can be two entirely different things. His practice of questioning the validity of an alleged divorce action as a central issue in cases even when the divorce is not in dispute by the litigants illustrates his recognition of the potential ambiguity of actions. In his attempt to determine the intent behind a particular action, Shaykh Hamid assessed the evidence that was presented in court in light of different possible meanings of actions by considering the potential scenarios in which the action could have taken place, which is similar to how Moroccan judges considered the "social identity" of individuals to assess the intent behind action (Rosen 1984, 1995b). For Shaykh Hamid, this required an understanding not only of the litigants and their actions, but rather the context in which the action took place and if another meaning could logically be ascribed to it.[3]

In the following three cases, Shaykh Hamid assessed intention by considering the different meanings of actions in the potential scenarios in which the action could have taken place. Then, he determined which meaning was most probable based on litigant and witness testimony. Cases were rarely brought to court with a "disputed divorce" as the central issue, and the questionable nature of a divorce usually arose later in the proceedings. Also, cases that involved a dispute about a divorce action were not always opened by women. Those that were opened by men most often involved a claim that a wife had left without a lawful reason. However, when the wife was summoned to court, she often argued that she left because she had been divorced by her husband, as we will see with Abdulmalik and Mariam.

A Wily Wife: *Abdulmalik v. Mariam*

Let us now return to the case introduced at the beginning of chapter one. The plaintiff was Abdulmalik, a shy, soft-spoken man of about 30 years old. He came to court in December 1999 to demand the return of his wife, Mariam. According to Abdulmalik, Mariam had left their home in the village to live in Zanzibar Town with her paternal aunt. She refused to return to him because she said she was divorced. Abdulmalik explained that one month earlier, he was late returning home after the evening prayer and Mariam was angry with him. She was well-educated in religious matters and was in the habit of giving him lessons in religion every night after the evening prayer. They went to bed, and when he woke up in the morning she asked him if he knew how a man lawfully divorces his wife. He replied that he did not know, and she told him to get a pen and paper and she would teach him. She told him to write the words "I, Abdulmalik, divorce Mariam, who will no longer be my wife, three times." Abdulmalik explained that he wrote the words, but that Mariam suddenly "turned everything around." She took the paper, told him she was now divorced, and left for town. After hearing his account, the *kadhi* and clerks recommended that he open a case. Mariam would be summoned to testify. As Bwana Fumu prepared the *madai,* he explained to Abdulmalik that, in his own opinion, he did not think the divorce would be ruled valid. He said he thought that because Abdulmalik had assumed he was getting a lesson in religion, there was no intention to enact a divorce in the action of writing the statement.

Madai

1. The plaintiff is a man, aged 30, from N—.
2. The defendant is a woman, aged 19, from N—.

3. The plaintiff and defendant have been married for the past two years and have one child.

4. The plaintiff claims that he usually goes to the mosque to pray in the evening and goes straight home afterward.

5. The plaintiff claims that the defendant is in the habit of teaching the plaintiff religious matters when the plaintiff returns from the mosque in the evenings after praying.

6. The plaintiff claims that approximately one month ago he was late returning home from the mosque and the defendant, his wife, was very angry with him.

7. The plaintiff claims that the next morning the defendant said to him "My husband, come here and I will teach you how a person divorces his wife; get a sheet of paper and a pen and write what I tell you to write." The defendant told him that he should write, "Mariam, you are no longer my wife," three times.

8. The plaintiff claims that the defendant then turned everything around and told her husband that he had just divorced her three times.

9. The plaintiff demands that the defendant return to his home if she is still his wife because the plaintiff did not intend to divorce her, it is only that the plaintiff feels that he was being given a lesson; he was merely being taught by his wife in their usual manner.

10. The reason for this suit is that the defendant claims she has been divorced three times by her husband, the plaintiff.

11. This claim originates from N—, Unguja, in the Northern A district.

12. The plaintiff asks the court to rule that:
    a. The defendant must return to her husband.
    b. The defendant must pay all court fees.

Mariam appeared in court about two weeks later. She was wearing a black *buibui* and a sheer headscarf, and had a chubby, gurgling baby with her. Mariam was rather sullen and seemed displeased to be in court. After Bwana Fumu read Abdulmalik's claim to her, she explained her side. Factually, their stories were very similar: Abdulmalik came home late on the night in question, and she was angry. She said he had told her he had been with his other wife. They went to bed, and she explained that the next morning she told Abdulmalik that she wanted a divorce and asked him to write it for her. She said that he agreed and wrote the paper to divorce her three times. Her counterclaim stated simply that she had been lawfully divorced three times and she wanted the court to recognize it as valid.

Majibu ya Madai

1. The defendant agrees with points 1–4 of the *madai*.
2. The defendant does not agree with the explanation of the plaintiff in point 5. She claims that the plaintiff divorced her, and that she had not been teaching him for about one year.
3. The plaintiff agrees with point 6.
4. Concerning points 7 and 8, the defendant explains that she asked her husband, "Where are you coming from? Why are you late today?" The plaintiff answered that he was coming from the home of his other wife. Indeed the defendant explained that she has dissolved her marriage lawfully and that she wanted her husband to write her the divorce. And the plaintiff indeed wrote her a divorce for three times.
5. The defendant does not agree with the *madai* in point 9, and she explains that this is because she has been divorced three times.
6. The defendant does not agree with points 10 and 11 of the *madai*.
7. The defendant asks the court to rule the following:
    a. The court refutes the *madai* of the plaintiff because the defendant has already been divorced three times.
    b. The plaintiff should pay court fees.

When Abdulmalik and Mariam gave their testimonies to the court, their accounts were much like the initial claim and counterclaim. The *kadhi* and clerks had few questions for Abdulmalik, but asked Mariam about her habit of instructing her husband in religious matters. On the day of the alleged divorce, had she told him what to say to divorce her as if she were teaching him? Had she told him that she was giving him a lesson? She answered "yes" to both questions. Abdulmalik's older brother had accompanied him to court on this day, and Shaykh Hamid reviewed the details of the case for him. The *kadhi* and the clerks found the situation highly amusing, and when Shaykh Hamid was reviewing the case, everyone except Abdulmalik and Mariam laughed. When addressing Abdulmalik, one of the clerks teasingly referred to Mariam as *mwalimu wako* (your teacher).

I interviewed Mariam a few days later. She had come to court to talk to the *kadhi*, but he had gone to Zanzibar Town to do some errands.

As in the initial hearing, Mariam seemed agitated and a bit unhappy. She shifted a great deal in her chair, and her eyes darted around the room. Despite this, she was quite forthcoming about her case. We talked about the divorce, and Mariam told me that she had asked Abdulmalik to divorce her, and that because he did not know how to do it she showed him. I asked her if she would agree to go back to Abdulmalik if it turned out she was not really divorced. She seemed puzzled by what I said, and so Bwana Fumu, who had been listening with interest to our conversation, explained to her that the *watu wa dini* (people of religion, meaning the *kadhi)* would look over the divorce paper and decide if it was valid. Bwana Fumu tried to assure her that Abdulmalik still wanted her to be his wife, but Mariam was clearly dismayed at the idea of going back to her husband, and insisted she would not go back to him because they had "had enough" of each other.

In their next meeting with the *kadhi,* it became clear that Mariam and Abdulmalik agreed about the events that had taken place, and there was no doubt in anyone's mind that Abdulmalik had written the divorce paper. The central issue was thus not whether a statement had been written, but what the statement meant. The action was not in dispute, but its meaning was: Mariam claimed that Abdulmalik was granting her request for a divorce, but Abdulmalik claimed he was merely receiving a lesson in religious law. Under these circumstances, had a valid divorce taken place? Shaykh Hamid told them that intention was the critical issue, and explained its importance using an example: if a person threw a rock and hit someone in the head, would it matter if the thrower had not intended to hit the victim? Shaykh Hamid pointed to the large stack of books on his desk and cheerfully told the litigants that he was working hard on the case.[4]

The following week, everyone was back in court, and Shaykh Hamid ruled that Mariam had not been validly divorced. He argued that this had been established by her own testimony: Mariam confirmed that she was more educated than her husband, and she admitted that she had been in the habit of giving him lessons in religion and had asked him to write the divorce paper. Therefore it was plausible that Abdulmalik thought he was getting a lesson when he wrote the divorce statement. Although the divorce had been written out properly, the words were not enough. To further this point, the *kadhi* asked Mariam if Abdulmalik had spoken the words when he wrote it. She said that he had not: she said them and he wrote them down. The *kadhi* used this as further evidence of Abdulmalik's lack of intention. He argued that it was credible that Abdulmalik had not engaged in the action of writing with the intention to divorce his wife because he could have legitimately interpreted the act of writing as a lesson, as he had claimed. Therefore, the action had no validity as a divorce, and Mariam was still his wife.

To Shaykh Hamid, the divorce action could have multiple meanings. In essence, Mariam's role as a household religious instructor established the likelihood that Abdulmalik had no intention to divorce his wife. The *kadhi* reasoned that because she played this role in their marriage, Abdulmalik assumed he was merely receiving a lesson. Although Mariam claimed she had asked him to divorce her, she admitted that he had not said the words and only wrote what she said. In his decision, Shaykh Hamid wrote that intention is necessary for a valid divorce by repudiation. He argued that the testimony of the husband that he did not intend to divorce his wife was substantiated by the wife's admission that she had in fact instructed him in matters of religion. Although Abdulmalik's character was not cited in the written judgment, it was evident through their laughter and jokes that the *kadhi* and clerks considered him a gullible man duped by a clever wife. This was one of the few cases in which Shaykh Hamid cited a legal source other than the Qur'an or *hadith*.[5]

> After listening to the claims of the plaintiff and defendant and investigating their case, the court sees that, truly, the plaintiff did not have intention to pronounce [the divorce]. Therefore the court rules that after swearing an oath, the section of the law in *Bughyat al-Mustarshidin* No. 372 applies: "*Only a divorce with niyya, intent, is suitable.*" The testimony of the defendant claims that she asked him for a divorce and she pronounced it as she taught him what to do. Therefore, by the testimony of the defendant and the relevant marriage laws the divorce is not proven. Therefore, the defendant Mariam is truly still the wife of the plaintiff Abdulmalik. And the court orders him to support her according to the law and if the plaintiff or defendant is not satisfied with this decision they have one month to appeal. This case is closed.[6]

The ruling in this case would be difficult to understand without a knowledge of religious education in Zanzibar, and Shaykh Hamid's decision can be read as his recognition of the significant role women can play in household education and their ability to achieve status as a learned individual. Shaykh Hamid believed there was more than one possible meaning of the divorce action based on Mariam's role as a religious educator in their home. Although Abdulmalik had studied until the sixth grade and never attended *chuo* (Qur'an school), Mariam was relatively well educated: she had studied until the eleventh grade and attended Qur'an school for a number of years. Abdulmalik emphasized that he was a religious man: he prayed every day, led the evening prayer at his local mosque, and was instructed in religious matters by his wife. He also told me that he had no idea why his wife would want a divorce since until the night he returned late there had been no animosity between them. When I interviewed Mariam, I told her that Abdulmalik said she was well educated and taught him in religious matters. She agreed. She said that she "knew a little" and that Abdulmalik had asked her to teach him. Although she claimed that she had not done it for a while because of her pregnancy, she said that she had taught him to divorce her, but only after she told him she wanted a divorce.

Education is highly valued in Zanzibari families for both boys and girls, although widespread school attendance is a fairly recent phenomenon. There were few schools in rural areas before independence and the 1964 revolution and formation of the United Republic of Tanzania. After the revolution, new schools were built and there was a significant increase in the number of students enrolled. For men and women over age 40, school attendance was much rarer, especially among women. Many older women told me that they did not receive any education—religious or secular—when they were growing up. Most men and some women were sent to Qur'an school, but only a few men in the rural areas attended the government-run colonial schools; no women of my acquaintance attended secular schools before the revolution. Among female interviewees over 50, any type of school attendance was rare and they reported that there was very little value placed on a secular education for girls and, in some families, boys. People cited a number of reasons for the lack of education. When discussing secular schools, they expressed elders' fears that children might be tempted to convert to Christianity. In terms of religious education, many people told me that it was a "time of ignorance" when people did not know that Islam encouraged lifelong education for boys and girls. Bi Nini, a woman of about 60, told me that only her brothers were put in the government school when they were children because "the elders of the past were stupid." She said the "girls only learned to sweep, wash dishes and raise children. Nothing else, not even the *chuo*." She explained that the *wazee* used to say that girls should not study in school. "But these days we understand," she said "Each girl goes to school and then when she gets home she goes to *chuo*."

Bi Kombo, in her early fifties, told me that when she was young it was rare for girls to study at all. She did not study even though her father was a Qur'an teacher. She told me, "In those days they [the girls] were told that if they were put in school they might become a *mhuni* (delinquent)." Bi Pili, in her sixties, said that the elders of the past thought that attending the colonial-run schools was sinful. She had studied in *chuo* for a short time, but said that it was one of her husbands who truly taught her to study the Qur'an. Some women said that their elders had thought the colonial schools were very dangerous, especially for girls. One elderly woman told me that the in Zanzibar's colonial period, elders were worried that the schools were teaching only Christianity and Christian ways, so they hesitated to send either boys or girls to the schools. This attitude also affected some younger women. Asia, Nini's daughter-in-law who was around 40, said that she was not put in government school because her father told her that it was *haramu* (forbidden) and therefore she would study only in the *chuo*; he told her that in the government schools one could not study *dua* (religious matters, prayers). Today, however, school attendance is very high, even in rural areas, and most youngsters attend both government schools and Qur'an schools. A recent World Health Organization study on children's health confirmed my suspicions about widespread government school attendance; the study found that 71 percent of all school age children in northern Unguja were enrolled and 91 percent of children over 12 years of age were enrolled in school (Montresor et al. 2001). Both boys and girls are encouraged by their families to make as much progress as possible through the government and, usually, Qur'an schools. Family expectations for a student's performance do not seem to differ based on gender, and in interviews from 2000 and 2002, one schoolteacher and two *kadhi*s informed me that girls generally performed better in school than boys.

As noted in chapter three, most people regard marriage as something that should follow the completion of one's education. A sad story told to me by a friend illustrates this. She told me about a girl from Pemba who was married off at age 18 when her parents heard that she failed her Form IV exams so could not continue in school. After she was married, however, it came to light that there had been a mistake and she had passed into Form V. Because married women cannot return to public school, it was a tragedy of lost opportunity both for the girl and her family, who could have benefited from her education and the possibility of gainful employment.

It is also quite common for adult women and men to continue or embark upon a religious education. Many adults make a sustained effort to continue their religious education through attending adult Qur'an schools, discussion circles, or lessons at home from a friend or family member (see also Purpura 1997). In my interviews, many men and women emphasized the role of family members, particularly children and spouses, as personal religious educators in their lives. One woman in her sixties who had been married and divorced three times, Bi Fatuma, stressed that all of her three husbands had been religious teachers and each had given her personal instruction, which she valued. She also noted the role of her grandchildren as current religious instructors, which was not uncommon for many older women. Bi Hawa, also in her sixties, told me that every evening her grandsons came by to teach her what they learned in *chuo* that day.

A Headstrong Husband: Makame and Mwajuma[7]

A few months before Abdulmalik opened his case, a man called Makame came to court, also claiming that his wife had left him without just cause. Makame was about 45 years old, and stood out from the other men by the socks and shoes he wore (most men wore rubber flip-flops or went barefoot), and his unusual manner. He was one of the few visitors to the court who seemed concerned with privacy, and he spoke in hushed tones when he explained to the *kadhi* and clerks that his wife Mwajuma had left him and returned to her family after claiming she was divorced. He said he had not "written anything" (meaning a divorce) nor had any desire or intent to divorce her. His claim, prepared with the help of the clerks, demanded that Mwajuma return to him. Mwajuma was summoned to court a few days later. She was a pretty, vivacious young woman of about 25 with an easy-going manner and a ready smile. When the clerks read Makame's *madai* to her, she responded by saying that she had already been divorced by *kulipa fedha* (paying money) to Makame. Her counterclaim called simply for the court to recognize the divorce and for Makame to pay all of the court costs.

When the litigants appeared together to give their testimonies, Makame talked at length about their life together and explained his many grievances against Mwajuma, which were listed on a piece of paper he brought with him. He spoke slowly and, referencing his notes, complained about her habit of leaving the house without his permission and her refusal to prepare food for him. In an attempt to hurry him along, the *kadhi* told Makame he only wanted to hear the details surrounding her departure from the marital home to return to her family. Makame answered that she had left four months ago and that, despite his pleas, she refused to come back. In her testimony, Mwajuma simply claimed that Makame was lying and that he had divorced her four months ago. When the *kadhi* asked how many times he had divorced her, she replied that it was not a repudiation but rather that her elders had "given him his money," meaning it was *khuluu*. She said that Makame had been given 45,000 Tanzanian shillings (about 60 USD) in front of witnesses and that he had then written a divorce paper. She produced the paper and showed to the *kadhi*.

Because the stories of the litigants were in direct opposition, the *kadhi* surmised that one of the litigants was lying and announced that they must return with witnesses the following week. Makame's witness was an elderly man named Juma. I found it odd that he was from Mwajuma's home village and one of her distant relatives. The *kadhi* asked him what he knew about the dispute, and Juma explained that he knew the couple well and that he had been asked to be a witness when Makame divorced Mwajuma! Makame was both shocked and distraught to realize that his own witness was testifying against him, but Juma went on to explain that Makame had divorced Mwajuma first through repudiation, and only later asked her for money when she refused to go back to him.[8] After some argument, Mwajuma's elders had eventually paid him the money, which showed that a *khuluu* divorce followed the initial repudiation. To clarify, the *kadhi* asked Juma if he had actually seen Makame receive the money and write the divorce paper; Juma answered "yes" to both.

Mwajuma's witness was another relative from the same village, Kassimu. He explained that he knew she had been divorced and that he himself had written it by proxy at Makame's request, since Makame was illiterate. The *kadhi* paid special attention to Kassimu's testimony and asked numerous questions about how he wrote it and if he wrote it for money; he was determined to establish whether this was lawful *khuluu* and not an incidence of "writing for money." Kassimu told him again that he wrote it specifically at Makame's request, and that it was for 45,000 shillings. As with the first witness, Makame was upset with Kassimu's testimony and interrupted repeatedly to ask if he had really been asked to write the divorce by proxy. Because of Makame's growing frustration and because both litigants said they had other witnesses, the *kadhi* agreed to hear more testimony, even though Mwajuma's version of the events had already been supported by the first two witnesses. Although Mwajuma's second witness, an uncle, was present and ready to testify that day, Makame claimed his witness was old and infirm and thus could not possibly come to court. After some discussion, the *kadhi* agreed to call on the witness at home the following week provided that Makame find a vehicle and gas money.

Two weeks later, the *kadhi*, Bwana Fumu, another clerk, and the litigants made a house call on Makame's witness, who was yet another of Mwajuma's relatives. As with many journeys involving a large group, there was some confusion when they left the court together, and the visit lasted about two hours. I did not go with them for reasons I do not recall; there probably were not enough seats in the borrowed car. When they returned to court the *kadhi* announced that the case would soon be finished and he scheduled a day when he would read the final judgment. I asked about the house call, and Bwana Fumu told me curtly, "Makame divorced her." Laughing, Shaykh Hamid added that after all Makame's effort to find another witness, "He merely said the same thing as the others—that Makame had indeed divorced his wife and taken money for it!" Although Shaykh Hamid was highly amused by the whole matter, Bwana Fumu and the normally cheerful Mwajumu were clearly annoyed that the case had wasted so much time by progressing this far.

The party returned the following week for the ruling. Shaykh Hamid explained that it had been established without a doubt that Makame had validly divorced his wife. He cited the testimony of numerous witnesses as evidence that he had written a divorce and received money for it. He made a specific point of establishing that the divorce was *khuluu* and cited the Qur'an and *hadith* literature on the practice of proper *khuluu*.[9] Makame was upset at the decision, and Shaykh Hamid explained to the litigants that they had the right to appeal the decision to the Chief *Kadhi* within 30 days. We later learned that Makame appealed, but the Chief *Kadhi* upheld the decision of the lower court.

This case is notable because of the emphasis Shaykh Hamid put on establishing exactly how the divorce occurred, and what Makame and Mwajuma's intentions were surrounding the circumstances of their divorce. Even after numerous witnesses had testified that Makame had indeed divorced Mwajuma, Shaykh Hamid sought to understand the precise circumstances surrounding the demand for money and whether the divorce was valid *khuluu*. As in other cases, Shaykh Hamid took particular care to establish how and why money changed hands in this situation to ensure that it was not a case of writing for money but was rather a proper *khuluu,* which should entail the wife's desire to end the marriage. Although Mwajuma and the witnesses never claimed that she asked him to divorce her, the *kadhi* did establish that Makame did not receive money for the divorce until after she was at her home village and refused to go back to him. Mwajuma's refusal to return to Makame was sufficient evidence of her agency in the situation (according to Shaykh Hamid this is necessary for *khuluu,* unlike repudiation) and her intention to remain divorced from her husband. I got to know Mwajuma quite well since she lived near me, and she told me that this was indeed what had happened—although she did not ask Makame for a divorce, she did refuse to go back to him. And although she claimed her elders were upset at his request for money, they eventually agreed to pay him. Establishing Makame's intentions in the matter was simpler. Many witnesses had seen him divorce his wife and receive money for it; no other explanations for the action were proffered, and Makame's denial of the divorce was simply considered a lie.

Leila's Contract: Leila and Abeid

Leila was a reserved woman of about 30 who came to court to sue for maintenance from her husband, Abeid. She was attractive and always carefully dressed and was one of only a few women who wore a black *buibui* and headscarf, rather than a *kanga,* to court. She told the clerks that Abeid, a boat-builder, had not maintained her properly for the past year. Although nothing in her *madai* suggested that she had been divorced, during the proceedings the central issue turned out to be whether a contractual action-driven divorce had taken place. This bitter, drawn-out case was the only one I witnessed that involved a conditional divorce. Shaykh Hamid explained conditional divorce to me as one in which a man divorces his wife by telling her that if she performs a certain action, like sitting in a particular chair, then she will be divorced. In Leila's case, however, the conditional divorce was set by her elders and herself. Because she was not getting adequate maintenance, her elders prepared a contract that avowed that Leila would be divorce if Abeid did not fulfill the terms within.

When we talked in court, Leila did not mention the divorce. She told me that she opened a case because from the early days of their marriage, her husband had not supported her adequately. She was the third of his four wives, and after their wedding he had wanted her to live with her cowives in a nearby village. At first she refused and stayed with her parents, but later she said that she "followed *sheria*" and went to live with them for the next five years.[10] When I asked her if *sheria* required a man to provide each wife with her own quarters, she said that this was true, but only if the husband had the means to provide it. She ultimately tired of living with the other wives, and left to live with her aunt in town. Although Abeid eventually told her that a house—her *own* house—was ready for her, she said she would not go back. She was angry with him for not supporting her, and was demanding back maintenance. The clerks helped her prepare the *madai,* which stated that Abeid had not supported her or their child for two years and demanded back maintenance at 1000 shillings (about 1.50 USD at the time) per day for the whole period for a total of 730,000 shillings. The claim stated that if he was unable to do this, she wanted a divorce.

Madai

1. The plaintiff is a woman aged 32 from S—.
2. The defendant is a man aged 40 from P—.
3. The plaintiff and defendant have been married for 13 years and have one child.
4. The plaintiff claims that the defendant has not maintained her or their child for the past two years; he has not provided food, soap, oil, or any of the proper marriage maintenance.
5. The plaintiff also claims that she has problems with the defendant because she has no place to live and is living at her family home.
6. The plaintiff's elders prepared a contract [which specified] that the defendant must support his wife and their child and build them a house, but he has not fulfilled it.
7. The basis of this claim is that the defendant does not maintain his child or the plaintiff and has not given them a house.
8. The claim originates from S—, Unguja.
9. The plaintiff begs the court to rule that:
   a. The defendant pay her back maintenance for food and the other items which he did not provide at the rate of 1,000 shillings per day for 730,000 shillings total.
   b. If he pays this amount, after the case is closed, he is requested to return to live with the plaintiff in a proper manner as husband and wife, and the defendant must maintain the wife in the normal manner and follow the contract prepared by the court.

    c. If the defendant fails to pay the maintenance, then he must give the plaintiff a divorce but must continue to support their child until he is able to support himself.

    d. The defendant must pay the court fees.

The stoutly built Abeid cut quite a figure when he came to court. He was a loud, assertive man of about 40. He was reputed to be fairly well-off, and this was reinforced by his expensive clothing, large belly, and heavy cologne. Abeid had a tendency to be both demanding and quarrelsome, and it was clear that the court staff found him difficult to work with. Bwana Fumu read the *madai* aloud, and Abeid disputed most of it. His counterclaim stated that he had been supporting Leila adequately before she left to live with her sister. He itemized the food and clothing he had given her, although it remained unclear when or how often he had done so. He also claimed that he had a house ready for her, and that she broke "the laws of Islam" by leaving him and refusing to return.

Majibu

1. The defendant agrees with points 1–3.
2. Concerning point 4, the defendant does not agree with the explanation in the *madai*, and he answers that the first time the plaintiff left him to go to her elder sister he left her food, like 10 kg of rice, 10 kg of flour, 5 kg of sugar, 1 gallon of oil, one fish, and other things. The third day the plaintiff explained "I am leaving and going to the city" and the plaintiff left that day, and the defendant left her much food and went to her sister and asked where she was. Until today, she has not returned to the defendant.
3. Concerning point 5, the defendant answers that he has already found a house for the plaintiff to live in, but she will not return to her husband the defendant because she is still living with her sister. She has broken the laws of Islam by refusing to return to him.
4. Concerning point 6, the defendant answers that he has no problems with the plaintiff's demand for maintenance, but that she left him contrary to the laws of marriage.
5. Concerning point 7, the defendant answers that he maintained his wife and their child like normal.
6. Concerning point 9a, b, c, and d, the defendant does not agree with these and answers that the problems have already been resolved by their elders.

7. The defendant begs the court to:
    a. Order the plaintiff to return to the defendant if he is indeed her husband.
    b. The plaintiff pays court fees.
    c. Other issues that might arise.

When they appeared in court together, Leila began her explanation by using a legalistic argument. She stated that Abeid had not been "fulfilling the laws of marriage." At first, he left her for two or three months and did not support her. She said that because she was not able to wait around for him, she went to live with her sister in town. She explained that she had taken the problem to her elders and that Abeid said he would pay them 50,000 shillings in back maintenance but had not done so. Leila finished her testimony by stating that he had four wives, implying that problems were to be expected with a man who married four wives, and that she was tired of these various problems and would not return to him. She was clearly exasperated when the *kadhi* assured her that the role of the court is to solve their problems, and she quietly explained that her elders had solved the problems long ago, but that now she was "tired of it all." It seemed that her patience had worn thin.

Abeid looked tense and annoyed throughout Leila's testimony; at times he almost seethed. When he was invited to speak, he explained that he had fulfilled all the duties of a husband. He admitted that he had left her, but only because he was called away to take care of an ailing relative. When the *kadhi* asked him if he supported her, Abeid replied angrily that he maintained her adequately with both food and clothing. It seemed that he then lost his ability to maintain any semblance of calm, and roared that it was indeed Leila and her elders who "broke the law" because she went to live in town without his permission! Trying to calm himself, he explained that he did not want any problems with Leila or her elders, but that he could not agree to divorce her because "there is nothing in Islam relevant to this." It should be noted that this mirrors Leila's use of a legal discourse to explain the trouble.

Two weeks later, Leila came back to court with her elders, who were to serve as witnesses; Abeid was also present. She was from a family well-known for religious learning, and one of her uncles was the same Mwalimu Simai who was introduced in chapter two. According to many people in the area, he outranked the *kadhi* in learning. Shaykh Hamid knew this, and treated the party with great *heshima* (respect) throughout the proceedings.[11] Leila's grandfather spoke first, and told the *kadhi* that he knew of the problems and that he had asked Abeid to give her a divorce but that he refused. Mwalimu Simai explained that they had all met together to try to solve the problems in the marriage. They told Abeid that he needed to support his wife adequately and either find or build her a house. They had dealt with the matter by writing a contract that stated if Abeid did not fulfill the terms, Leila would be automatically divorced. They called this the *karatasi ya wazee* (elders' paper) and said that it stated that Abeid would pay 50,000 shillings in back maintenance to Leila's elders and, if he did not, then she would be divorced. They said that although Abeid was not present at the time of writing, he had agreed to the terms. The *kadhi* then turned to Abeid to ask if he had indeed agreed to the terms. Abeid answered that he had, but argued that even though that he had not fulfilled the terms of the contract, Leila was not validly divorced.

Noting this, Shaykh Hamid asked the group if a time period, *muda*, had been set for Abeid's payment. The elders said they had given him two weeks. Abeid tried to contest this by saying that he had been under the impression that there was no time limit, but ultimately his protest did not matter because in the ensuing commentary of the elders it became clear that Abeid had indeed returned to the village on the determined day to give Leila the money. However, she was in town that day, and was thus not there to receive it. By the time she returned from town, Abeid had already gone. At this, Leila nodded, and said she reasoned that she was divorced because he had not fulfilled the contract: she had not received the money by the date specified in the elders' paper.

Abeid was very annoyed with the proceedings and the presence of the witnesses, but Shaykh Hamid calmly informed him that he would have a chance to bring his own the next time they met. Abeid replied by shouting, "*I'm ready!*" He then bellowed dramatically that he thought everyone in the courtroom was a "liar" and that there was "no *sheria* present." Everyone else remained calm, and Shaykh Hamid coolly apologized to Leila's witnesses for the disturbance.

The case resumed two days later. Abeid's witness was a middle-aged man who said he knew that some "mistakes" had been made in the marriage, but that Abeid had never divorced his wife. The witness explained that Abeid had four wives and treated them equally; using a legal discourse, he added hypothetically that a man must provide for each of his wives, and he commits a sin if he fails to do so for a year. He also mentioned the contract, but seemed to think that the amount specified was a gift to Leila rather than a payment of back maintenance.

The *kadhi* and the witness talked for a long while, and their conversation eventually centered on whether the contractual divorce had actually happened; they seemed to agree that a complication was presented by the fact that Leila was absent when Abeid arrived with the money. The witness argued that the problems stemmed from Leila's absence and Abeid's confusion about whom he should pay since she was not there.

At this, Shaykh Hamid told Abeid that he had erred and had perhaps broken the contract because he did not go to Leila's father with the money when he realized Leila was not present (recall that the elders' said the contract required him to take the money to her elders in the first place). The *kadhi* continued by saying that he was challenged by the situation, "We have come to hard place." He reviewed, and said that "One person said he did not bid farewell, another says he did; one person says he left food, another says he did not." At this point, however, he indicated that the case rested on the validity of the contract. Was Leila divorced? If Abeid had not paid the money by the end of the *muda,* then Leila would have been divorced. Abeid, however, said that he came back to her village to pay her in a timely manner. Because Leila was not there when he came, her family assumed she was divorced. However, there must have been some question about the validity of the divorce because she came to court to sue for maintenance and, clearly, Abeid never accepted that she was divorced.

On the final day in court, only Leila and Abeid were present. The *kadhi* announced that he had prepared his decision and would read it to them after reviewing the case to demonstrate how he had reached his conclusion. Abeid was, as usual, loud and argumentative and he interrupted the *kadhi* often. Leila was quiet and rather morose, but smirked at many of Abeid's comments; it was hard to imagine that they could successfully reconcile. In Shaykh Hamid's review of the case, it was clear that assessing Abeid's intentions was an important part of his reasoning. Shaykh Hamid explained that Abeid had not maintained his wife for an entire year, and had erred by leaving her without telling her. He reviewed how Leila's elders solved the problem by writing a contract requiring Abeid to pay back maintenance by a certain date or Leila would be divorced. He noted, however, that it was in fact *Leila* who broke the contract by being absent when Abeid came with the money. Abeid's intention was thus to pay: he made a good faith effort to do so, and even slept at the village for two days awaiting her return. Because Leila did not appear, she broke the contract, and no divorce had taken place. Shaykh Hamid told Leila that he had explained his reasoning to her elders, and they had agreed that she had not been divorced and believed that she should return to her husband.

This was not the end of the matter, however, because the question of Abeid's failure to pay maintenance had to be resolved. The *kadhi* summarized Abeid's offenses and ruled that he must pay her back maintenance for 10 months at 400 shillings per day, for a total of 120,000 shillings, which was quite a bit less than Leila's *madai* requested. He added that the case would only be closed when this money had been given to Leila.

Hukumu

The court gives a judgment according to the laws of marriage. The defendant must take his wife and support her fully according to the laws of marriage. Then, for the time period which he did not support his wife he must pay her 400 shillings for each day for a total of 120,000 shillings; this is ransom for the plaintiff's food. This case will be finished and if either the plaintiff or the defendant is not satisfied then they have one month to appeal. If they do not do so in this time period, then it will be completely finished.

Discussion

These cases show that Shaykh Hamid determined intention by considering the social context of specific actions, and the possible scenarios that could lead to those actions. As we have seen, assessing intentionality requires knowledge not only of the litigants and their actions, but also the context in which the action took place and if another meaning could be ascribed to it. The cases underscore the necessity of contextualizing legal interpretations. In Abdulmalik's case, Shaykh Hamid's decision could not be understood without recognition of the social role of religious education and educators in Zanzibar, where it is indeed reasonable to assume that a woman can serve as a religious educator in a marriage and thus have religious authority. Makame's case was similar. The *kadhi* considered the possible scenarios in which Mwajuma and her family could have paid Makame for the divorce. His focus on intention was important in establishing what kind of divorce it was, and if it was a valid *khuluu*. Through witness testimony, Shaykh Hamid determined that Mwajuma herself desired a divorce because she refused to return to him after the first instance of divorce: her intention to remain divorced legitimated the exchange of money as *khuluu*. No other explanation for the actions surrounding the divorce was proposed, and the testimony of Mwajuma and numerous witnesses established that Makame had indeed divorced his wife, and intended to do so by first issuing a divorce and then requesting money in *khuluu* in front of many observers. In Leila's case, Shaykh Hamid reasoned that even though Abeid had not actually met the terms set out in elders' paper, he had intended to do so by going to the village with Leila's money at the prescribed time. Thus, because the intention to fulfill the contract was there it was in effect fulfilled and Leila was not conditionally divorced.

Messick has discussed the development of the "shari'ah subject" in recent Yemeni history through an analysis of legal literature (1998, 2001). He explores how jurists of the Shi'i Zaydi school have approached intentionality in their writings, and argues that the individual shari'ah subject, as presented, was essentially "constituted by intent" (2001). However, there were significant differences in the meaning of intention between reciprocal acts (e.g., contracts of sale) and unilateral acts, like divorce by repudiation. According to some jurists, expressions can have "independent weight" in the words alone, but to others, a divorce cannot occur without intention (2001). One Yemeni jurist called al-'Ansi proposed that while "unambiguous" expressions have inherent meaning, linguistic analysis is necessary to determine intention in "indirect" expressions—these are words like "go home!" that could indirectly indicate divorce. Such a statement is not necessarily binding, and thus a connection between "expression" and "meaning" (intent) must be established (170). Another thinker, Shawkani, opposed such analysis and proposed that divorce by repudiation happens whenever it is intended. Along the same lines, an unambiguous statement of divorce that a man claims he uttered without intent cannot be considered valid since only the husband knows his own intent (2001: 172).

Theorists in some Islamic legal schools suggest that intention is not as significant an issue in divorce as in other areas of law. John Esposito has noted, for example, that in Hanafi law a divorce pronouncement is considered final even if uttered in jest because of the serious nature of divorce (1982). This view implies that the action of making the pronouncement is more significant than the intention behind the pronouncement. Laleh Bakhtiar notes that the Maliki and Shafi'i traditions agreed with Abu Hanifa, and that the famous scholar Ibn Rushd (also known as Averroes) wrote that according to Abu Hanifa and Imam Shafi'i, intention is not necessary in divorce (1996; citing Ibn Rushd's *Bidayat al Mujtahid*). However, we have seen that for Shaykh Hamid, who works in the Shafi'i tradition like all Zanzibari *kadhi*s today, intention is the foundation of validity in divorce. In Abdulmalik's case, he cited a collection of *fatwa*s from Hadrami Shafi'i mufti as requiring intention in divorce, which suggests differing opinions among Shafi'i scholars on intention.

Messick's analysis of the reasoning of the Zaydi thinkers provides an interesting comparison to Shaykh Hamid's reasoning. Although on the surface Shaykh Hamid's position seems similar to that of Shawkani, a more appropriate parallel can be drawn to al-'Ansi's discussion of indirect expression. This is because of the way in which Shaykh Hamid views the written divorce: in Abdulmalik's case, Shaykh Hamid emphasized that he had not pronounced the words in conjunction with writing, further demonstrating that the written words were not necessarily a divorce action. We can therefore consider Abdulmalik's written statement an indirect expression to which the meaning must be attached. From the beginning, Abdulmalik denied that he intended to divorce his wife, but his word (even with an oath) was not sufficient proof of this, and Shaykh Hamid did further investigation into what actually happened.

How then can meaning be attached to an indirect expression? In Zanzibar, Shaykh Hamid supposed that intent could be determined, and his approach is similar to what Rosen found in Morocco and among Moroccan judges, where assessing the inner states of others relies on understanding social context and personal circumstance, and "background knowledge and knowledge of another's mind are intertwined" (1984: 56). Shaykh Hamid did not normally emphasize determining the character of the litigants through the witness testimony, but rather assessed the evidence presented by the litigants to determine the context of the contested action. Because actions are often ambiguous, the role of the judge was to see the possible meanings of a certain action or statement, and to decide which was most probable based on litigant and witness testimony. Assessing intention did not require an intuitive ability to read the mind of other individuals, but an ability to perceive the scenarios in which the action could have taken place. In conjunction with chapter three, then, we see that for Shaykh Hamid, intention and proper action must go together: actions that indicated divorce do not constitute divorce without evidence of the statement of repudiation, and a repudiation without intent also does not make a valid divorce.

This emphasis on the context of a particular action is similar to what Peter Just found among the Dou Donggo of Indonesia, a place far removed from Zanzibar. Just argues that fairness and liability are not based on notions of intentionality and causation, but rather on the consequences of certain legal actions, and on the entire social context of the actor who performed the misdeed (1990, 2001). Furthermore, for the Dou Donggo, individuals are not held responsible only for those actions they committed as is established through allegations and testimony, but also for those they would have been likely to commit. Although Shaykh Hamid did not hold individuals responsible for the potentiality of their actions, both types of reasoning involve an imagining of the possible scenarios involving the actor and the legal actions in question. In his work on the Sherpa of Tibet, R.A. Paul has proposed that in general theoretical terms, "schema theory" is helpful in understanding how intention is assessed cross-culturally. Through culturally constructed schemata, individuals learn cultural scripts (Shank and Abelson 1977) that are typical of certain situations. Then, "By contextualizing a single moment within a larger script or scenario and drawing on clues about character and circumstance, one can usually guess, with only an occasional surprise, which of the possible scripts is operative at any given moment" (Paul 1995: 18). This is similar to Shaykh Hamid's assessment of intention.[12] Actions are ambiguous, and it is the role of the judge to determine the meaning. The potential ambiguity of actions is indicated by the way in which Shaykh Hamid often framed the validity of alleged divorces as the central issue in cases—even when the divorce is not disputed by the litigants. I propose, then, that the "scripts" upon which Shaykh Hamid draws are dependent on historical circumstances, in that they take into account changing gender relationships and norms of behavior. In Abdulmalik's case, Shaykh Hamid's understanding of Mariam's role as religious educator and the couple's pattern of evening religious instruction suggested a logical alternative for the meaning of writing a divorce that supported Abdulmalik's denial of his intention to divorce, which outweighed Mariam's claim that she told her husband that she wanted a divorce. To come to his decision, Shaykh Hamid took into account current norms of education in Zanzibar for someone of Mariam's age. I suspect he (or another *kadhi*) would have decided the case differently 50 years ago, or if the woman in question was 50 years older than young Mariam. As noted earlier, my interviews indicate that far fewer women were receiving religious (or other) education in past decades and thus it would be less likely that an older woman would be instructing her husband in religious matters. Hence, it would be less likely that Shaykh Hamid would reason that a woman of that age would have duped her husband into writing a divorce by offering to instruct him.

In cases like those presented in this chapter, the *kadhi's* stress on intention elucidates local ideas about male and female agency and motive in marriage and divorce acts. In both cases, the *kadhi* considers current norms of gender relationships in marriage to aid his decision-making process. In Zanzibar, as is likely the case elsewhere, divorce by repudiation or through compensation is not always a straightforward process. The cases in this chapter show the role women play in determining their own marital status. Here, I build on the work of contemporary scholars of Islamic law that seek to demonstrate how women participate as active agents in a seemingly male-privileged legal tradition (Tucker 1998, Mir-Hosseini 1993, Hirsch 1994, Fair 2000). Shaykh Hamid considered women to be agents in their own marital fate. In Makame's case, once Shaykh Hamid accepted that the divorce did take place, he made a particular effort to note what was behind the exchange of money. Knowing that many Zanzibari women are the victims of unlawful "writing for money," the *kadhi* sought to establish that Mwajuma did indeed desire a divorce from her husband. Although she claimed she did not ask him to divorce her, the fact that she returned to her natal home and refused to go back with him when he asked were indicative of her intention to be divorced and her agency in the matter. This reasoning is consistent with the *kadhi's* view of proper *khuluu* divorce.

In Abdulmalik's and Mariam's case, Shaykh Hamid might appear to be a patriarchal *kadhi* siding with the man in an ambiguous situation. However, a more nuanced understanding comes from considering his judgment with respect to the social role of religious education in Zanzibar. Mariam's agency is recognized by the court. Although it was not written as such in the judgment, the *kadhi* and his clerks both assumed that the clever Mariam, who tutored her husband in religion, had tricked him into giving her a divorce. Even though the ruling was against the wife, the decision shows that the *kadhi* believes in a woman's ability to engender her own divorce through pressuring her husband. Fair (2001) and Middleton (1992) have mentioned that Swahili women will provoke their husbands to divorce them through a campaign of irritating behavior. Though I do not know of any woman who admitted to doing this, it probably happens with some regularity.

In Leila's case, a number of important issues come to light. In addition to Shaykh Hamid's determination that Abeid's intention to give the money to Leila outweighed the failure to actually do so, we must take into consideration knowledge and authority. In Griffiths' work on dispute resolution in Botswana (1999), one of the primary aims is to consider the strategies women choose in making claims on their male partners in a variety of legal institutions. Importantly, she notes that men and women pursue different legal strategies. Relationships are essential in structuring the way litigants conceive of their problems and the avenues they can take in search of a settlement or satisfaction. Social status, as it is determined through kin and family networks, influences decisions about court use, legal strategy, and presentation of claims. Therefore, variations in women's socioeconomic status affects their litigation strategies. Through her meta-argument that social networks must be considered in thinking about how people, especially women, learn about legal forums and pursue legal options, we can understand a bit more about Leila's case. Hers was the only one I witnessed in which the family had drawn up a divorce contract. This was likely because Leila came from a family and a village well known for religious learning. Griffiths clearly shows that women pursue different marriage strategies and legal strategies based on their position within households of the community (45).

CHAPTER FIVE

Witness and Authority: Elders, Shehas, and Shaykhs

Shaykh Hamid was not the only voice of authority in the Mkokotoni court, and in this chapter, I look at the practice of witnessing to explore the way in which the *kadhi* utilized other modes of authority in the process of resolving disputes. The practice of witnessing had a dual nature in Shaykh Hamid's court. Through giving testimony, a witness (*mshahidi*, pl. *washahidi*) produced evidence. Also, as a procedural step, witnessing reflected state, community, and familial webs of power and influence. As we will see, the *kadhi* made a point of incorporating both local authority (elders and religious scholars) and state authority (government appointed *shehas*) into procedure. These different modes of authority were emphasized in the way in which people describe the process of resolving marital disputes. Both lay people and legal professionals emphasized a tripartite method that was typically described as follows: first, a disputing couple takes the matter to their *wazee* (elders); if the elders are unable to resolve the problem, then disputants visit the *sheha*; if the *sheha* fails, then the dispute is taken to the *kadhi's* court. As we will see, the practice of witnessing in court reflected this process, since it was the parties who had attempted to resolve the dispute who were most often called as witnesses.

The first time I met Shaykh Hamid, he followed this general outline when describing dispute resolution and, like his fellow *kadhis*, he described his work as the final step in the process if consultation with elders and *shehas* failed. He specifically noted the importance of elders in their role as witnesses to disputes, and as the people who should know what was truly going on in a marriage: "Our religion says that if a wife and husband argue, then they need to call two elders, the *mzee* of the woman and the *mzee* of the man, in order for them to solve the problem...this is *sheria za dini.*" When a disputing couple takes the first step by going to their elders, he said, they should be prepared to act as witnesses. If the mistake was on the part of the husband, then the wife's parents should be able to act as witnesses against him. If the mistake was on the part of the wife, then the husband's parents should act as witnesses against her. Although he also described the importance of *sheha*s in the process, he explained that since they did not usually know a great deal about religious law, they tended to mediate disputes through counseling instead.

Elders play perhaps the most important role in dispute resolution, and people often cited the Qur'anic condition for a couple to seek counsel from older family members.[1] However, in interviews, many people indicated that elders played a much greater role in the past when people were less likely to go to a *kadhi*. I was often told that elders used to handle nearly all disputes, and it was only rarely that they were taken to an extra-familial authority like a *kadhi* or *sheha*. People often commented that elders were quite successful at solving most marital problems, although this was often attributed to their frequent use of the *bakora* (beating stick) on their children—even adult children. Mzee Bweni, in his mid-sixties, told me that if a woman had problems with her husband, then all the *wazee* would get together to discuss the problem. Whichever party had no *adabu* (manners) was considered responsible for the problems, and then, "[he or she] would get five whacks with the *bakora* and everything would be fine. This was our *mila* (custom)." Mzee Bweni explained that now people used the *bakora* less often, which he attributed to increasing study and knowledge about religion: people had learned that beating is not appropriate *sheria za dini*. "Nowadays," he said somewhat regretfully, "people are not shy to go off to the *sheha* or the court with their problems." The reason for his wistful tone came clear another day, when he told me that he disapproved of airing one's dirty laundry in public at the court. You will recall that Mzee Bweni was also the man who thought the *kadhi* was *chongoo* (one-eyed) and biased in favor of women. Mzee Chumu, about the same age, told me that people in the past were more likely to seek divorces from their elders. When I asked if they used *sheria* in such matters, he answered,

> Yes, they knew it [*sheria*]. But young people these days go to court or *sheha*. Me, I've never gone to the court or to the police. Today, it is like a game for them, but in the past, we just fixed it ourselves

in our own neighborhoods. There were *kadhi*s, but we were able to resolve our differences there at home.

When I asked why things had changed, he simply said, "I don't know. Maybe because in the past, if you were the troublemaker your *wazee* would beat you with a stick three times! Bam!" He explained that "today's behavior" stemmed from the fact that elders no longer beat their children when they caused problems that threatened the marriage. "In the old days," he continued, "If you were beaten by your elders you were afraid and didn't behave foolishly again. But me, I married and never went to my elders with a problem, or even the *kadhi wa mtaa* (local *kadhi*)."

Although "local *kadhi*s" may also play a significant role in resolving marital disputes, they do not have the official role that *sheha*s do. *Sheha*s have long had a role in local level dispute resolution (see also Ingrams 1931) and are now a mandatory step on the way to court: before opening a case, claimants must present a referral letter from a *sheha* stating that he or she tried to resolve the problem. My interviews suggest that the office and role of *sheha*s has changed dramatically since the revolution; most people told me that in the past, *sheha*s were chosen from within their communities and were widely respected. Although this may be a somewhat idealized view of the past since it seems that for at least in the past century, *sheha* appointments required approval from higher authorities, the sentiment does reflect current political tensions and the disregard many people have for *sheha*s. As discussed in chapter one, in the nineteenth century it was likely that the *mwenyi mkuu* had authority to appoint *sheha*s, though this may have changed after the last *mwenyi mkuu* died in 1865 (Ingrams 1931). Middleton and Campbell write that in the colonial period, *sheha*s were selected by the public, but were subject to the approval of the District Commissioner (1964: 45), and Stockretier notes that *sheha*s were regular mediators and witnesses in colonial-era *kadhi*s courts in Zanzibar Town (2008: 80).

Today, *sheha* authority certainly comes from above since the government appoints them. *Sheha* appointments are announced over the radio, which serves as notice to both the appointee and to the community. In my interviews, all *sheha*s told me that the appointment could not really be refused. After the 1964 revolution, ruling party "chairmen," who presided over political districts called "branches," replaced *sheha*s briefly. *Sheha*s were reinstated by Act No. 11 of 1992, which renamed the branches *shehia*s and designated *sheha*s as the "chief government executives" of the *shehia*s; they were to be appointed by the Regional Commissioner in consultation with the District Commissioner.[2] The act sets out the numerous qualifications for the post: a *sheha* must be at least 50 years old, literate, and come from the *shehia* over which he or she presides.[3] Among the other duties specified in the act, *sheha*s are responsible for the "settlement of all social and family disputes arising in that area in accordance with the customary laws of that area."[4] Although a *sheha* cannot be a leader of a political party, he or she may be a member and they have normally been members of the ruling party, *Chama cha Mapinduzi* (CCM). In northern Unguja, most people view *sheha*s as political appointments who are highly loyal members of the ruling party. Correspondingly, some *sheha*s are viewed with skepticism and distrust, particularly by those sympathetic to the opposition party.

In 1999–2000, I interviewed each of the 11 *sheha*s who presided over the *shehia*s that were closest to Mkokotoni (this is perhaps half the number of *sheha*s in the northern A district of Unguja). At the time, all were male, though by 2005 a female *sheha* presided over Mnazi Mrefu, the *shehia* in which I lived. All of the *sheha*s told me that their primary task was to solve the problems of people in their communities. Although they are not trained in law and receive only minimal instruction when appointed, they are nominally responsible for all types of disputes; none claimed any expertise in religious or secular law, and most said that they settled problems according to community norms. The *sheha*s were consistent in their assertion that it was the job of the *kadhi,* not the *sheha,* to use religious law. Despite this unanimity, the *sheha*s differed when assessing their own abilities to handle disputes. The *sheha* of Kibazi, for example, explained that there were two types of problems he faced: problems of marriage (*kindoa*) and problems of the community (*kimtaa*). He said that he could and would only deal with the latter, and therefore sent all marital problems directly to the *kadhi.* As he saw it, *kindoa* matters necessarily concerned religious law and were thus beyond his capabilities, and not part of his duty as *sheha.*

> If we get problems concerning marriage and divorce, we send them to the *kadhi.* We write a letter for them, because as *sheha*s we solve problems of *kimtaa.* For marriage, we will call in the elders and try to solve it, but if the [disputants] won't agree to the marriage, then we send them over there, because we don't solve matters of *kindoa* here. So, if there is a marriage case, we'll send it over there [to the court] with a paper saying something like "these two

don't get along, and they are fighting"... it is not really our job to resolve marital disputes.

Other *shehas*, however, were more reluctant to send such matters to the *kadhi*. The *sheha* of Bandarini told me that he disliked sending problems to the *kadhi* simply because he preferred to handle them on his own within the community. Accordingly, although he presided over the largest *shehia* in northern Unguja with 10,000 residents, it was only rarely that disputes from his *shehia* made their way to the court.

These views illustrate two different interpretations of what marital disputes entail. Those *shehas* who considered any marital dispute to be a legal matter were quick to send their disputants to the *kadhi*. The others thought that marital problems could be resolved outside of court through counseling and drawing on local norms of marital behavior. The *sheha* of Mwembeni, for example, explained that he was able to solve many marital disputes because, as a local, he was familiar with the *utamaduni* (culture) of the area.

> I haven't studied law, but I'm used to this kind of leadership and I know the culture of this area. I was born here and have lived here for a long time. Thus, I'm never overwhelmed by the customs of this place—I understand and recognize what people are doing, and I understand that if I do such-and-such I can resolve many problems.... When a *sheha* is given his position, he is given the "inside" of his own village and he knows the *mambo* (matters) of the area and the habits of the people there.[5]

Local *kadhi*s, *kadhi wa mtaa*, may also play an important role in marital disputes. Although the local *kadhi* was never cited in descriptions of the three-part process of dispute resolution, several people told me that they should be consulted before going to court. Unlike the *sheha* whose authority comes from the state, the influence of local *kadhi*s comes from within the community: their status comes from popular recognition of their expertise in religious scholarship, particularly law. Lay people often seek out such individuals to contract marriages, certify divorces, settle inheritance disputes, and answer legal questions. It was not unusual for Shaykh Hamid to ask a prominent local *kadhi* to assist with complex cases, and he sometimes deferred to such expert opinions. As we have seen, this figure was most often Mwalimu Simai, who was considered by many people—Shaykh Hamid among them—to be his superior in religious learning; Shaykh Hamid once told me that when he called Mwalimu Simai it was as a "special witness." I knew Mwalimu Simai fairly well, and since he lived nearby we had the chance to talk on several occasions; he was a charming and engaging conversationalist and I enjoyed our talks. When I asked him about his role in dispute resolution, he explained that he worked alongside the *kadhi* and *sheha*s, but suggested that he and the *sheha* did parallel work since they both attempted to resolve problems before a matter would be taken to the state-appointed *kadhi*. He noted that sometimes people took to their problems to the *sheha* and sometimes they came to him. Modestly, however, he implied that people only came to him when the *sheha* could not help them: "he [*sheha*] might tell them to go to someone to get a better understanding or explanation of their problem." Indeed, I recall one case in which a young woman explained that she had first gone to her *sheha,* but he had directed her to the *shaykh,* by whom he meant Mwalimu Simai. Similarly, Mwalimu Simai told me that he often advised people who came for his help to also visit the *sheha* for another perspective on the problem. He told me that there were many local *kadhi*s, and emphasized the importance of seeking help from different sources, "If the person you ask is not able to fix your problem or help you out, then you go to another place . . . you'll go by one, then you'll go by another." This echoed what he told me in another conversation, when he said that all Zanzibar *kadhi*s should consider all *madhhab*s when handling cases: he thought that the best solution to a problem could be found only by considering many opinions.

Interviews suggested that taking marital problems to the *kadhi*s and *sheha*s is more common today than in the past, and many people were ambivalent about the change in resolving disputes. Many simply noted that "times change." A few men, like Mzee Bweni, viewed the change negatively. Women, however, sometimes remarked that the change benefitted women because they now felt comfortable going to the *kadhi* for a divorce, whereas in the past they would have been encouraged to persevere in the marriage. This is similar to what Hirsch has noted about changing perceptions of *kadhi*s' courts among Swahili women in Kenya (1998: 82). Bi Nuru, a woman of 60 who had been married and divorced several times, held this view. She was married the first time at a very young age, and at the time of our interview she was having problems with her fourth husband. She told me that in her second marriage, she and her husband fought a great deal and so she went to the *sheha:* "It was just the *sheha,* but the *sheha* of the past . . . you didn't go to the court back then." Her parents accompanied her, and the *sheha* called in her husband, who claimed Nuru was lying about their marital troubles. She said they all believed him, then beat her with a stick and told her to go back to him: "Women had no rights. I was beaten in front of everyone!" Today, she said, things were better because in the past, "women had no one to speak for them." Now, she and other women listen to the radio and were much more aware of their rights and thus had more options; she said that, in her opinion, women today were not shy about going to the court.

Bi Mboja, in her forties, expressed a somewhat different view. She had been married and divorced once. Her husband had not maintained her or their children and had not provided them with a place to live, but refused to divorce her when she asked him to do so. When I asked her why he would not, she explained that divorce creates trouble and anxiety. Eventually, however, she said that she went to her elders (noting that this was the first step to take), who told her to persevere in the marriage. She did so and waited a long time, but the situation did not improve and he still refused to divorce here. As a result, she went to the *sheha* and to her husband's parents to tell them that she wanted a divorce. One of her husband's relatives finally convinced him to divorce her through writing a statement of repudiation. Mboja never went to court, and though she said that she thought some women might be ashamed to do so, her own reasoning was different: she said she hesitated because the situation and her feelings could have changed rapidly, and she might have regretted taking such concrete action as seeking a court-ordered divorce:

> The government would have given you the divorce because he did not want to divorce you. Then you've already been divorced, but if your husband returns, that's it. He can't return [to you].... So I decided that I should just wait and get my divorce by his hand, not the hand of the government.

How and When Shaykh Hamid Called Witnesses

The 1985 *Kadhi*'s Act specifies rules of procedure for the *kadhis*' courts. Section 7 requires that witnesses in all *kadhis*' courts, including that of the Chief *Kadhi,* be called and heard without regard to religion or gender, and that facts be assessed based on the credibility of witness testimony rather than on the number of witnesses.[6] This is a significant alteration of classical Islamic laws of witnessing, in which a man's testimony is equivalent to that of two women, and the testimony of a

Muslim is more valuable than non-Muslim testimony. However, other than a general requirement that a witness know something about the dispute, Shaykh Hamid rarely specified who or what type of person (male, female, Muslim, non-Muslim) should be brought as a witness. Initially, I assumed this was because of the *Kadhi's* Act. However, he once told me he was unaware of the restrictions placed on witnessing by the act. Another time, he explained that both state and Islamic *sheria* required witnesses to be called, but told me that two women must be called to equal the testimony of one man, and did not reference the act. Although Shaykh Hamid thus did not consider the act directly relevant to his court procedure at that time, I did not see him differentiate between witnesses on the basis of gender. Although he often asked litigants to bring a father or father's brother as a witness, he never indicated that female testimony would be less valuable, and it was only once that I heard Shaykh Hamid tell a litigant specifically that she should bring a male witness.

Shaykh Hamid said that, ideally, witness testimony should be heard whenever possible. On the practical plane, however, he usually called them only when the essential facts of the case were in dispute, as we have seen in previous chapters, and thus ruled on several cases without calling witnesses. The type of witnesses he called to court reflects the tripartite process of dispute resolution, since it was most often *shehas* and elders who came to court to give testimony. The *kadhi* usually spent a great deal of time explaining the process of witnessing to litigants. Those litigants who were in court for the first time were often unclear about who should be called and what would happen when they came to court. Shaykh Hamid distinguished between the testimony of a *sheha* and the testimony of ordinary witnesses. Because *shehas* are an official part of the dispute resolution process, they were expected to have some knowledge of the disputes and were often called in even when ordinary witnesses were not deemed necessary. Litigants, both male and female, most often brought family members, specifically elders, as witnesses.[7] Shaykh Hamid was somewhat ambivalent about the role of elders as witnesses. He told me once or twice that a person's elders were not ideal because they were not neutral, but he also often specified to litigants that elders were the best possible witnesses because only they could really know what was going on in a troubled marriage. Regardless, it was elders who were most often brought to court. As we will see in the following cases, elders were not called in only to give testimony, but as his aides in resolving the dispute and as moral guides in the lives of their adult children.

Witnesses were nearly always called in maintenance disputes because they normally involved a disagreement about whether a man was supporting his wife adequately; witnesses were called to provide testimony on what food and clothing he had provided for her. As we saw in chapter one, the largest category of cases opened by women concerned maintenance. When they came to court, women told the clerks, "*sipati sheria zangu*" (I'm not getting my legal due). Even women with little education know that *sheria* requires their husbands to maintain wives and children fully: he must provide a house, food, and clothing. Although women are expected to provide the *vyombo* and cook, clean, and tend children, men do not generally open cases against wives who do not perform these duties. It is possible that men brought such complaints to court, but highly unlikely that the clerks would have agreed to open a case with that as the central complaint. When men occasionally complained about wives who were lax in these duties, Shaykh Hamid always responded by explaining that if a woman did not want to do household chores, then it was a husband's responsibility to do them himself or hire someone:

> Because of religious law, you have to build your wife a house in which she can live well, and you have to feed her well. Then you give her household implements, clothes, take care of her…don't espouse any hatred or bitterness. You can't force her to do anything. You have no right to force her. But some people, they make their wives do things, and they don't have the right. You see these women getting firewood, getting water, and cooking the food. You see women doing this all the time, but it is not the law. They are helping out, but it isn't the law. The law says that you [the man] have to do it all. The husband has to fulfill everything! But she might work if we agree to it together, this is not religious law—if they agree together is *mila*.[8]

In Kenya, Hirsch found that most complaints brought by women to *kadhis*' courts similarly concerned maintenance. In a notable contrast, however, men came to court claiming "rights" to their wives' labor and elders usually supported a man's claim to a wife's work (1998: 99). Hirsch notes that when asked, *kadhis* responded that Islamic law does not require women to work. She writes that men were often surprised that they had no legal recourse to demand a wife's labor, but emphasized local norms of gendered household contributions even after learning this.

The cases that I describe as "maintenance claims" are those in which maintenance was the primary claim in either the plaintiff's initial complaint or in the *madai*. For example, if a woman claimed maintenance with the addendum that she wanted a divorce if her husband fails to support her, I categorized it as a maintenance claim because support was the primary aim. In the court register, Bwana Fumu and the other clerks wrote entries in a formulaic way, and almost all maintenance claims were written as *anadai chakula, nguo, na pahali pa kukaa hana* (she demands food and clothing, and has no place to live). Maintenance claims varied significantly, however, and ranged from women requesting more food on a daily basis from an otherwise adequate husband, to women requesting back maintenance for months or years from absent husbands. Also, since it was not unusual for cases to go through many manifestations, a woman might come to court requesting a divorce or registered divorce but to have her claim eventually dealt with as a claim for improved maintenance, as we saw in previous chapters.

In ruling on these cases, Shaykh Hamid took into account a number of different factors when he decided a man owed maintenance. Although the baseline figure for daily maintenance was 800 shillings (just over 1 USD at the time), he considered factors like the financial status of the husband (but not of the wife), the number of people he must support, and whether the wife was at fault in marital discord. Also, if a husband was ordered to pay back maintenance for a period when he did not support his wife, the length of time was taken into account. If it was long, as in many months or years, the prescribed per diem was lowered. Although many women were unhappy with this, it seemed to be to ensure at least a portion of the payment would reach her; if the amount was too high, a husband might simply flee.[9]

Shaykhs and *Sheha*s in Court

Two cases introduced in chapter two are fine examples of the roles of *sheha*s and local *kadhi*s in the court. In one, *Mosa v. Juma*, we see how the *sheha's* official role was incorporated into *masharti*. In this case, Mosa argued that her husband Juma did not provide adequate maintenance for herself and her children, some of whom were from a previous marriage; she also claimed that he had no house for her to live in. Juma claimed that he supported her properly. Because there was a disagreement about maintenance, Shaykh Hamid called the *sheha* and witnesses, and Mosa and Juma brought two witnesses each. Eventually, Shaykh

Hamid issued terms of reconciliation, which specified that Mosa should return to Juma, Juma should support Mosa, and either should see the *sheha* with any future problems. In my earlier discussion of the case, I noted that Shaykh Hamid explained that she had erred because she did not return to Juma and did not go to the *sheha* to report the continuing problems with him. He told both litigants that their legal responsibilities included observing the *masharti and* reporting to the *sheha* if they had problems. Thus, the law had been violated by breaking the terms and by neglecting the *sheha.* In the end, Mosa "bought" her divorce in part because she failed to see the *sheha* as ordered in the ruling.

The case of Machano and Aisha shows the potential influence of a local *kadhi* in the court. Recall that Shaykh Hamid introduced Mwalimu Simai to the litigation differently than he did regular witnesses. He did not emphasize his role as a witness, and did not use the term *mshahidi* for him because of his status as a religious authority; we saw similar respect paid to Mwalimu Simai in Leila's case, discussed in chapter four. Instead, he emphasized Mwalimu Simai's role as his aide—he was there to help solve the problems of the couple. Furthermore, although ordinary witnesses were always required to swear an oath, those who were called to help "solve" the case rather than simply give testimony were rarely required to do so. Shaykh Hamid's quick dismissal of the *edda* problem indicated the weight of Mwalimu Simai's testimony as a religious authority: he considered the question of *edda* settled because a noted religious authority said that he had lawfully married the couple, and he was not going to contest Mwalimu Simai's authority to do so. This appears to be similar to the way Moroccan judges consider the "situated individual." Rosen has argued that a judge will assess witness testimony based on his knowledge of the witnesses, how he or she is socially "situated," and his or her character in the social realm (2000: 73). As with litigants, the credibility of the witness is established through his or her social networks, and the truth of a statement is determined by assessing the truthfulness of the one who made the statement. Social context is thus all-important, and witness testimony thus creates facts. The reliable witness produces reliable testimony that can be considered fact, just as we saw Mwalimu Simai's testimony settle the *edda* question immediately.

A Procedural Error: *Hamza v. Kombo*

The first case we will take up in this chapter involved yet another dispute over whether a divorce had occurred and illustrates the importance

Shaykh Hamid places on proper procedure and the necessity of calling witnesses. The case dragged on for four months, and was frustrating for all involved, primarily because it concerned a previously closed case. It all began when Hamza, a stout man with a fluffy beard, came to court claiming that his wife, Kombo, had left him unlawfully a few years previously. After much discussion, the *kadhi* and clerks learned that Hamza wanted to reopen a divorce case filed by Kombo several years past. Although she had received a court-ordered divorce, they suspected there might have been a procedural problem because the former *kadhi* had never called witnesses, and this was noted in the *madai*.

Madai

1. The plaintiff is a man aged 36.
2. The defendant is a woman aged 25.
3. The plaintiff and defendant have been married for 10 years, and have 3 children.
4. The plaintiff claims that his wife has been prevented from returning to him for three years at her home.
5. This claim references case 8/97 in which the current defendant was the plaintiff.
6. The plaintiff claims that the court gave the defendant a divorce; after much research, it was apparent that there was a mistake in procedure and to fix this error the *kadhi* has decided to open another case.
7. The plaintiff asks the court:
   a. to listen to the counsel concerning the Chief *Kadhi* to make an appropriate decision.

When she responded to the *madai*, Kombo claimed that Hamza had not been maintaining her, and that she had been given a divorce by the former *kadhi*. In the *majibu*, she seemed to agree with everything Hamza stated in the *madai*, and was willing to have the earlier case investigated. However, because there were several people in court the day she responded, Bwana Fumu took her outside to prepare the *majibu*, and I did not hear their conversation.

Majibu ya Madai

1. The defendant agrees with points 1–3 of the claim, but states that her age is 29, not 25.

2. Concerning point 4, the defendant agrees that she has been home for three years, but that she got a paper of divorce from Mkokotoni.
3. Concerning point 5, the defendant says that it is true that this claim is in reference to case 8/97 in which the defendant was the plaintiff.
4. Concerning point 6, the defendant agrees that there was a problem with the [former] case because the plaintiff, who was then the defendant, never came to court.
5. Concerning point 7, the defendant agrees that the case should be heard and an appropriate ruling given.

Even though this case was unusual, it proceeded as normal, and the litigants came to court together to give oral testimony and to review what had happened all those years ago. They had a great deal to say; Hamza talked at length about Kombo leaving him, and Kombo argued that he had failed to support her for four months. All the while, Shaykh Hamid tried to determine if the previous *kadhi* had followed proper procedure. Eventually, the *kadhi* explained that they would need to call witnesses, and specified that the witnesses should be people who knew the details of their marriage and could offer testimony about what had happened. Hamza did not care for this suggestion, however, and said that they had "no witnesses." For some peculiar reason, he seemed to think that the case was finished, and declared that he was tired and had had enough. The *kadhi* told him, "Don't be tired!" and explained that they would not be finished until they reached a point at which both parties were satisfied. He instructed them one last time to bring witnesses, and they left.

The next time in court, Hamza announced proudly that he had no witness, but wanted to continue with the case that day anyway. By this point, we had all realized he had a flair for drama. Shaykh Hamid patiently explained that they could not continue with the case until he brought a witness, "If there is no witness, there is no one to prove whether things are true or not, and we can't drive the case forward." Somewhat stubbornly, Hamza replied that he would be very happy if they could proceed that day because he did not think he would come back to court again. Shaykh Hamid was frustrated, but maintained composure and repeated that although they could not proceed without witnesses, it would be fine to wait for a witness. Hamza said again that he had no witness, and the *kadhi* tried to suggest possibilities. They went back and forth for some time, and finally Shaykh Hamid told him that if he had no witness, then to bring any elder—mother, father, or anyone else who was still alive. Hamza shook his head slowly, said that no one would come, and mumbled solemnly about the difficulty involved in bringing a witness to court. At this point, the discussion of the witnesses had continued for over 30 minutes. Visibly irritated by now, Shaykh Hamid turned to Bwana Fumu and asked, "We've already listened to the case, but we're waiting for witnesses. We haven't got them, but the husband wants to finish the case anyway. What do we do?" Bwana Fumu, standing next the *kadhi's* desk, explained yet again the importance of witnesses to Hamza, and again suggested bringing an elder. The stubborn Hamza continued lamenting that "no one knew anything" and "no one would come." Finally, the *kadhi* offered to send one of the clerks to Hamza's village to find three witnesses: the *sheha* and two elders, one representing each litigant. Hamza accepted this plan, but before he left the room, he moved to the *kadhi's* side and asked about the divorce in a low voice, apparently suggesting that he wanted Shaykh Hamid to rule without hearing any more testimony. Shaykh Hamid replied loudly that in *his* court they used the process of summons and he could not rule on the case without hearing witnesses testimony.

A few weeks later, Hamza and Kombo appeared with witnesses and their *sheha.* Adding yet another difficulty, however, the *sheha* would not stay. He declared that he found the summons letter problematic: only one letter had been delivered to the *shehia,* but several names were listed in it. The *sheha* indicated that he wanted a letter addressed specifically to him, and that he would not return until he had one. Fortunately, the *kadhi* had more success with the others. Kombo's witness, a paternal uncle, knew the situation well. He explained that several years ago, Kombo opened a case because Hamza was not supporting her. The previous *kadhi* had ordered Hamza to pay 60,000 shillings in back maintenance, but because he had never answered his summons, the *kadhi* agreed to give Kombo a divorce. About a year later, when another man proposed to her, Hamza protested because he had not known she was divorced. Thus, he came to Shaykh Hamid, who had sent him to the Chief *Kadhi,* apparently to get clarification on the court-ordered divorce. The Chief *Kadhi* sent the matter back to Mkokotoni, and here we were.

During this explanation, Hamza began to get worked up. At the end, the *kadhi* told him he could ask the witness three questions. Instead, however, he asked the *kadhi* about the previous ruling requiring him to pay Kombo. The *kadhi* told him that he was supposed to ask the witness a question—not the *kadhi,* "Don't ask me the questions, my work is to write up the case!" The situation escalated—Hamza did not want to question the witness, and Shaykh Hamid became very aggravated with asking him to do so over and over. Hamza grew angrier and angrier, and Shaykh Hamid called in a police officer to stand by; the officer simply told Hamza not to make noise, and to ask his questions carefully. Peace somewhat restored, Shaykh Hamid then turned to Hamza's witness, whom Hamza made a point of identifying as "simply an *mzee* (elder), not a witness." Shaykh Hamid ignored this and asked the elder to comment on case. The old man answered that he had not heard anything about the case and had nothing to contribute to the proceedings. Shaykh Hamid ended the session by telling the entire party that they would meet again another day when the *sheha* could testify.

The next time the party was in court, as we were waiting for the *sheha* to arrive, Shaykh Hamid explained to me that he was trying to determine whether the previous divorce was valid. He and the clerks had perused the old case file, and he could not find any letters calling witnesses or any indication that witness testimony had been heard. I asked if the summons letters would have been put in the file, and he said that they would have, and showed me other summons. He surmised that two mistakes had been made: Hamza had never been informed of the divorce, and no witnesses had been called.

When the *sheha* arrived, the *kadhi* carefully explained his role: he was called to court because a *sheha* is both *mwenye watu* (the steward of the people) and their representative. He said that sometimes the *sheha* knows about the case, and sometimes he does not. Shaykh Hamid then reviewed the key elements of the dispute, and asked him who had caused the problems in the marriage. The *sheha* seemed quite pleased to be in court this time. He settled comfortably into his chair, and told the *kadhi* that it would be a long tale. The *kadhi,* who normally tolerated lengthy reports, was clearly wearying of the many hours spent on the case, told him to be brief and discouraged the forecasted long tale by asking simply, "Who was the *mkosa?*" (one who erred). Heeding the appeal, the *sheha* was brief. He said that Kombo and Hamza had both caused problems in the marriage: all those years ago, Kombo left without telling anyone, and Hamza had failed to support her for four months. To his knowledge, only one summons letter was delivered to Hamza in the former case.

The *kadhi* then asked the couple to leave so he could talk with the elders privately. When they left, he told the elders that the "young ones" were troubled, and that they all needed to talk together to solve the problems. The *kadhi* summarized the case in detail, and indicated the procedural problems in the case opened by Kombo years ago. He also expressed concern that Hamza had not answered the summons. I was confused, and asked if a wife could be divorced by the court if a husband did not answer a summons. Shaykh Hamid said that this was appropriate when the husband could not be found, but in this case Hamza had been right here on Unguja, and so the former *kadhi* should have sent the police to bring him to court. The elders listened intently, but did not add much to his reasoning process, and he did not expect them to.

Shaykh Hamid wrote the ruling a few days later. He noted that Hamza's failure to maintain his wife for four months was a neglect of his marital duties, but also that Kombo "forgave" him. This was unusual phrasing since the mistakes she was forgiving were not specified, and I was never clear why the phrase was included. The *kadhi* ordered Kombo return to Hamza, and ordered him to pay back maintenance of 102,000 shillings, the total of four months at 800 per day. If he failed to pay, Kombo would be divorced through *fasikhi*. Although Shaykh Hamid determined that the divorce issued by the other *kadhi* was invalid, he did not include any reference to it.[10]

Hukumu

After listening to the claims of the plaintiff and defendant and listening to their witnesses, [and] after the court's research it has been discovered that the fault is with the plaintiff. He stopped supporting the defendant according to her *sheria* [legal due] for a long time during the course of three years from July 25, 1997 until March 22, 2000.

The plaintiff assumed he could accuse his wife. The defendant forgives the mistakes of the plaintiff that were explained; he is her husband of three years, and she agreed that his mistakes were over [the course of] four months.

The plaintiff acknowledges that he failed to meet his obligations to his wife for four months. And all of the witnesses established that he failed her for four months.

Therefore, after research, the court rules that this has been proven. The defendant will return to her husband. The plaintiff will pay reparations to his wife from the days that he failed to maintain her. This is four months at 800 shillings per day and the total sum that will be brought to the court is 102,000 shillings.

If the plaintiff is not able to pay the reparations, then the defendant has the right to end her marriage according to the laws of marriage concerning *fasikhi*.

> *The wife is permitted to abrogate the marriage to her husband, [whether he is] present or absent, [if] it is unfeasible to acquire the spousal support in its entirety in three days and that determination is by the judge with witnesses or his own knowledge.*[11]

When the reparations are paid, then the case will be closed. If the plaintiff and defendant are not satisfied then they can appeal the case within one month. The court read the decision aloud. On April 25, the plaintiff will come to pay the money.

Although the case was complicated, it is a fine example of the importance Shaykh Hamid put on the testimony of witnesses. This was evident in his insistence that Hamza bring a witness and in his decision that the previous divorce was invalid because of procedural irregularities, one of which was the failure to call witnesses to the alleged lack of maintenance.

An Indecisive Wife: *Mpaji v. Jafari*

The next case illustrates the importance of elders as witnesses to a dispute. The plaintiff was a tall, robustly built woman in her late thirties named Mpaji who was not shy about expressing herself in the courtroom. Her vigorous persona, however, contrasted sharply with her attire. Mpaji was poorly dressed, even for rural standards. She was barefoot, and was wearing a faded and mismatched *kanga*—she wore one pattern over her head and shoulders, and another wrapped around her waist.[12] She came to court to open a case against her husband, Jafari, because he had not been maintaining her for the past 11 months. She said that she had left her husband's home, but explained that it was not because of poor support, but rather because she was ill and needed treatment. Indeed, throughout the hearing, she emphasized that she told him she was leaving (using the verb *kuaga),* as was proper, and thus had not violated her marital duties.

When she and Jafari came to court together, Mpaji testified that he had refused to let her go to her parents' home when she was sick and that he told her if she left, she should take her *vyombo* with her. Because removing the *vyombo* is an important signifier of divorce, as we saw in chapter three, she used this to indicate that Jafari had essentially threatened her with divorce. Mpaji was a powerful speaker and had a way with words, and it was hard not to be compelled by her narrative. When she continued, she said that she waited three days for his permission to leave. He did not grant it, so she left, taking two children and some *vyombo*. Jafari had not maintained her while she was away, and though he visited once after six months, she was gone at the time and did not see him. After several more months, she went to his elders to ask why he failed to support her. Next, she went to the *sheha*, who wanted to talk with her father about the marriage problems. She was very animated and often waved her large hands to emphasize a point, and it was clear that she was extremely frustrated with Jafari.

Shaykh Hamid invited Jafari to respond, but he had no questions for her. He then calmly explained that he had done nothing wrong, and that she left him of her own will. Quiet in both tone and movement, Jafari was quite unlike his wife. He said that when she left, he told her not to take the small children, but she had done so anyway. He had sent food, but she refused to return to him. Mpaji, who had been listening intently, avowed that she knew he had sent food, but she had not been home to receive it. She also said that although he had sent her money, the 5000 shillings was not enough for an entire month. To illustrate, she listed the various household items that she needed, and repeated again that the money was not enough to buy it all. Jafari countered that he had sent more than 5000 shillings, which sparked a heated quarrel. No longer placid, Jafari argued that Mpaji failed to tell him she was leaving, which led to another row about whether she had permission to leave. Shaykh Hamid listened for a few minutes, but quickly tired of the argument and announced that it seemed that neither party was getting their *haki* (rights). Hence, they must call in the elders.

When they were back a few days later, I noticed that Mpaji was wearing the same striped green dress, again covered with faded, mismatched *kanga*s. They brought their fathers as witnesses, and Shaykh Hamid introduced them to the case formally. He explained that cases proceed in a certain fashion, and part of the process was calling witnesses. Because this case concerned a marriage, he said, it was necessary to call the elders. He embarked on a lengthy explanation of why this was so, and focused primarily on the fact that no one else could really know the problems of the couple. He commented that the litigants had behaved rather badly, and then invited them to speak.

Mpaji's father, Mzee Ali, explained that she had been distraught when she came home, but told him that she left with her husband's goodwill. He said that Jafari had gone to Dar es Salaam on the mainland at some point, and that although he was supposed to have sent for Mpaji he had not done so. Mzee Ali did not know whether Jafari had ever brought food, but he knew that he had gone to see her once. She was not home that day, however, because her father had sent her to an *mganga* (natural healer or local doctor) in town for treatment for her illness, to whom he paid a hefty 20,000 shillings. The *kadhi* asked if Jafari had known his wife was sick, and Mzee Ali confirmed that he did. Jafari's father, Mzee Juma, then told the *kadhi* somewhat crossly that Mpaji had wanted a divorce for a long time, but his son had not agreed to it. He said that she had been sick, then left her husband, and never returned. This prompted Shaykh Hamid to ask the litigants if they wanted a divorce. Mpaji looked bewildered, but responded quietly "yes" with none of her usual bravado. Jafari said he had no desire for a divorce.

As with the previous case, Shaykh Hamid asked them both to leave and requested that the elders stay to discuss the problems. Shaykh Hamid reminded the men that it was their duty to raise their children properly and teach them about marital responsibility. Both men were attentive and were pleased to be consulted, and Mzee Juma contributed many of his own thoughts on the importance of parental responsibility. When Shaykh Hamid continued, he said that marriage was indeed difficult, and it was hard for elders to know exactly what happens "inside the home" of their children. However, since the *wazee* know more than others, they are normally called in to testify. Together, they reviewed the facts of the case. Shaykh Hamid said, "We know two things: she doesn't want him, but he doesn't want to divorce her." Under these circumstances, he said, a woman must buy her divorce in *khuluu*. However, he explained, the matter was complicated because Jafari had allegedly not supported her for 11 months. The elders agreed that the situation was difficult, but did not add much else, and Shaykh Hamid had not yet decided what to do.

The entire party was back one week later, and the case ended after nearly three hours of discussion. After the *kadhi* reviewed the details, Mzee Ali indicated that he had come to court with 100,000 shillings to buy Mpaji's divorce. It was clear that he had thought carefully about the *kadhi*'s explanation of *khuluu*: a woman must buy her divorce if her husband would not agree to it. Shaykh Hamid told me later that Mpaji and her father had discussed the matter at home and she had decided on divorce. This seemed a reasonable solution to all, but everyone was shocked when Mpaji suddenly announced, somewhat bashfully, that she still wanted her husband!

Mzee Ali was clearly exasperated with his daughter for changing her mind, but because of this new development, Shaykh Hamid closed the case with ruling of *masharti*. They would stay married, and Jafari would pay back maintenance. Although Jafari had provided some support, it was established that it was not enough for the maintenance of a woman and two children, and Shaykh Hamid wrote that Jafari would pay Mpaji 300 shillings per day for the 11 months. This was less than the *kadhi's* baseline figure of 800 shillings per day because the man was poor and because he had so many months to pay. The *kadhi* argued that the essential issues in this case were the lack of maintenance and the fact of her illness, which her husband—not her father—should have been paying to treat. Although the question of whether Mpaji left properly was important,. she had been ill, which was an adequate reason to leave without permission or a proper farewell. Shaykh Hamid stated repeatedly that Jafari was responsible for taking care of her, and because he had not done so, he had made a significant mistake and was liable for the expense.

Hukumu

After listening to the plaintiff and the defendant and their claims, the court decided that the plaintiff did not unlawfully leave her husband; rather she left to get treatment for an illness.

Concerning the problem of food maintenance, the litigants and their elders proved that the plaintiff was not receiving adequate food for a long time.

And the defendant did not support the plaintiff in her medical treatment for the illness with which she was stricken. The defendant stated that he did not do so because she was away at her parental home without lawfully leaving; according to the law the husband is the caretaker of the wife; he must provide food, clothing, household goods and the laws of the home. The husband does not have the right to argue with his wife, especially if he has not divorced her; therefore the court follows the verse of the Qur'an, which reads: *Men are the support of women.*[13]

In light of this verse, the court decides that the defendant must support his wife with food for the period of one year, each day 300 shillings for a total of 109,500. This indeed is the decision as it is given. If the plaintiff or defendant is not satisfied with this decision, then they have the right to appeal this case within a one-month period. This case is closed after giving this decision. The plaintiff should return to her husband, and her husband should give the plaintiff 10,000 shillings by the fifth month of the Islamic calendar.

His decision appeared to offend both litigants. Mpaji was upset that Jafari had to pay so little, and he was upset at paying so much. Mpaji then changed her mind again, and brought up the possibility of buying her divorce. Everyone was upset, and the entire party started discussing the appropriate amount for a *khuluu.* The clerks shuffled them out of the courtroom, however, because it was no longer a matter for the *kadhi.* When we discussed the case, Shaykh Hamid laughed about how difficult it was, and said he did not know what they would end up doing. Maybe one of them would appeal his decision, he said, or maybe Mpaji would buy her divorce. They never came back to court when I was there.

Elders and Witnesses in a No-Argument
Case: *Fatuma v. Omari*

Although Shaykh Hamid normally called witnesses only when facts were in dispute, there were some exceptions to this, as in the next case. The plaintiff, Fatuma, was a cheerful and energetic woman of about 30. I already knew her slightly because she was the sister-in-law of Mwajuma (from chapter four), with whom I'd become friends. Like Mwajuma, Fatuma was very personable and I was surprised and distressed to see her in court. She had come because her husband of a few years, Omari, was not maintaining her and their children properly. She said that Omari was a fisherman, and that she herself had no other work than farming and occasionally going to the beach with friends to collect the tiny shellfish used in soups and stews.

Omari was summoned, but this was only the first of many attempts to get him to court. Because men may spend days on fishing boats, his work led to many scheduling conflicts, and he was rarely available. When he finally came in, Omari agreed that he had not been supporting Fatuma adequately. He was a round-faced man who was just as pleasant as his wife. Although there was no conflicting testimony that usually necessitated calling witnesses, Shaykh Hamid decided their testimony was required anyway to determine the amount of back maintenance Omari should pay.

On their next day in court, Fatuma's elder brother testified as her primary witness; she explained that because she had no father, her brother was her *mzee*. Her brother said that Omari had left her for two years, but he did not know specifically what his transgressions were or how he had failed to maintain her. Shaykh Hamid was somewhat vexed that he did not know details of the case, and lectured the entire court on the importance of witnesses in maintenance cases—especially witnesses who knew something about the case. He explained that the court was like a hospital: it solved people's legal problems the way the hospital solved their health problems and cured their illnesses. Of course, the court could not do this without witnesses, and he stressed again the importance of bringing in someone who knew something about the matters at hand. He seemed to have given up on Fatuma's brother, and so he turned to her and Omari. Fatuma explained again that Omari had "run away" and did not support her during this time. The *kadhi* asked Omari how many days he was gone, and when he did not have a concrete answer, Shaykh Hamid berated the group yet again about the importance of bringing in people who knew something about the case. Because of this, he decided to postpone the hearing. He instructed Fatuma to bring an elderly man of his acquaintance, Mzee Machano, whom he thought could shed some light on the situation. He announced that they would all return the next day, and commented ir- ritably that "We did not get what they were looking for yesterday, and we did not get it today."

He excused the party, but they were slow to leave. Omar timidly told the *kadhi* that it would be difficult to come to court the next day. He added that he thought Fatuma's brother was "mixing things up," though it was not clear to what he was referring. Shaykh Hamid rebuked him loudly for leaving his wife and not giving her food, and told him to remind his elders that they would be continuing the case the next day. Shaykh Hamid immediately seemed to regret scolding the gentle Omari, and tried to ease his other concern by asking Fatuma's brother again about maintenance. The brother simply added that he knew Omari was not supporting Fatuma because she had come to ask him for money.

The next day the party returned with three elderly male witnesses, and Shaykh Hamid arranged to talk to them alone, without Fatuma and Omari. He told them that they were there to help solve the problem of the young people, and explained that Jabu had told him about her difficulties with Omari. He said that after hearing their testimonies, it was obvious that Omari was the problem, and that they needed to figure out what to do. "The husband has three *sheria* (here, duties)," he said, "and all are serious." First, he must pay his wife what he owes her for maintenance. Second, the husband and wife must agree to get along and try to reconcile to avoid divorce. Third, if the husband fails to support her, then the wife will get a *fasikhi* divorce. After this, Shaykh Hamid gave his usual lengthy lecture on the role of elders in their children's lives. Together, he said, he and the elders would ensure that the young people followed the terms of the court, and would reprimand them if they did not. As elders, their responsibility was to uphold *sheria* and to help their children—even adult children with offspring of their own—observe it.

Fatuma and Omari were invited back inside and Shaykh Hamid explained that even before he talked with the elders he knew who had caused the problems: Omari had failed his wife and his marriage in many ways. He had neglected his responsibility as the "leader" of the marriage and the head of the household. Shaykh Hamid told him that he had only one wife, and that she should be considered *ndugu*—a part of his family and a close relative. He continued by telling Omari that he should have gone to his elders for help if he could not support his wife, even though it was not their responsibility to support her. The *kadhi* explained that Omari would get a ruling ordering him to support her, and if he failed again, she would be divorced. Shaykh Hamid emphasized this last point by explaining that even if he failed to support her for only three days, she would get her divorce. The *kadhi* was in a professorial mood, and used this opportunity for a prolonged address on rights and duties in a marriage. He emphasized that a wife does not have the responsibility to provide anything in a marriage, and that she does not even need to cook if she does not want to, "This is the law, and we, as married people, are meant to satisfy each other and help each other."

Discussion

In his view of the duties and obligations of state-appointed Zanzibari *kadhis*, Shaykh Hamid recognized that his position was one of compromise. He was pleased to be a *kadhi*, but told me on numerous occasions that many others passed up the position because the Islamic courts have jurisdiction only over family matters, and thus "they [*kadhis*] would not be able to apply Islamic law in full." Shaykh Hamid seemed to view the compromised position of the *kadhi* as one that entailed obligations to many. This view of the compromised *kadhi* was evident in the way that Shaykh Hamid incorporated other types of authority into courtroom practice. *Sheha*s play an official role in dispute resolution, and Shaykh Hamid and the clerks upheld this by requiring litigants to bring letters from *sheha*s before opening cases. As we have seen, he often called *sheha*s in as witnesses, and in cases like Mosa's, included the authority of the *sheha* in his rulings by requiring litigants to take further problems to the *sheha* before returning to the court.

Although many people told me that couples were now more likely to go to the *kadhi* or the *sheha* than simply settling disputes with elders as in the past, it is clear that elders remain an important part of resolving marital disputes. In court, Shaykh Hamid regularly incorporated them not simply to give testimony, but as nominal advisors to him and as special moral guides for their adult children. He avoided using the term *mshahidi* (witness) with those advanced in age or learning. Instead, he introduced them to the proceedings as peers who were there to help the younger generation solve their problems. Although elders rarely informed the *kadhi*'s reasoning in a significant way, they were nearly always eager to help and eager to listen. As we have seen, when he excused litigants from the courtroom so he could speak privately with elders, he stressed their continuing responsibility in their children's lives and their duty to uphold *sheria* for the sake of the young. Fatuma's case shows the importance Shaykh Hamid places on the role of elders not just as knowledgeable witnesses about the goings on of the litigant's problems, but also on their continual role as supporters of the marriage, upholders of law, and exemplars for their adult children.

Not all *kadhi*s took this approach, however. Shaykh Faki, who worked in town, criticized his rural counterparts for this perceived accommodation of local authority. He distinguished between his decisions and those of what he called *shamba* (rural) *kadhi*s on the grounds that the local people intimidated the latter. Shaykh Faki worked in town when I knew him, but had worked for a time in a rural court, to which he commuted from town. When I wondered aloud if *shamba kadhi*s tried to get along with local people because they were from those very communities, Shaykh Faki agreed that the rural *kadhi*s strived to respect the elders. However, he thought it inappropriate because *kadhi*s are only supposed to "preserve people's rights." He said that he himself did not listen to local people, even though it caused difficulties during his tenure at the rural court:

> I didn't agree to listen to [local] people like this, and someone who disliked me hit me with a truck up there. I was leaving court... and I got on my Vespa to come home, and when I got to Madonde

someone hit me with a truck. Oh yes, I was injured and people thought I would die. Then, the Chief Judge decided it was better to remove me from there and move me here [to the town court]. But I absolutely did not agree [to listen to those local people].[14]

Zanzibari *kadhi*s may be appointed to courts in their home communities or elsewhere. Hirsch has commented on the significance of court assignment in Kenya, and writes that the fact that Kenyan *kadhi*s are usually assigned to courts outside their home communities affects their relationship with the communities they serve (1998). She proposed that Kenyan *kadhi*s may feel pressure from community elders to conform to local norms upholding patriarchal practices rather than interpretations of Islamic law that contradict such practice (1994: 218), and Karen Brison made similar observations in Papua New Guinea, where young magistrates often sided with women to "show their authority over older men" (1999: 81). Although I never heard elders make specific demands of Shaykh Hamid, he made a point of inviting them to assist in "solving" the problems brought to court, and treated local *kadhi*s with great respect. He encouraged the voices of the community in proceedings and his court provides an important contrast to what Hirsch observed in Kenya, where *kadhi*s and local elders seemed to exist in a contest over interpretations and applications of the law.

Recent scholarship suggests the importance of studying courts as arenas for the negotiation of social relationships, community-state relationships, and different normative orders (e.g., Starr 1990, Lazarus-Black and Hirsch 1994, Reiter 1997, Wurth 1997). Some scholars have considered the role of the court as a place of social drama (Gibbs 1963, Stoeltje 2002) and others have looked at courts and dispute resolution processes as arenas in which to affirm or contest community values and politics (Lazarus-Black and Hirsch 1994). In her work on how rural communities accept institutions and ideas from the state, Brison finds that rural magistrates view themselves as intermediaries between the local community and the state's new legal policies and programs (1999). She contends that the village courts are a valuable way to study competing visions of the nation because of the different ways state and local groups view and make use of the courts. Brison argues that one magistrate showed greater concern with defining a new polity through the new laws than with actually resolving conflict, thus developing his position as intermediary between local and state.

Shaykh Vuai, one of the other *kadhi*s with whom I worked, was one who emphasized the primacy of Islamic law in his court. Shaykh Vuai was based in town and traveled to work in rural courts; thus he was not part of the community he served, and was not connected by social ties in the way that Shaykh Hamid was. Like Shaykh Hamid, he knew his jurisdiction was circumscribed by the state, but he described his primary obligation as upholding the proper interpretation of Islamic law, even if this contradicted state legislation. Shaykh Vuai said he paid little heed to local or state authority other than accepting the *sheha*'s official role in dispute resolution (though I noticed that he was far less likely to incorporate *sheha*s into court proceedings than was Shaykh Hamid). Shaykh Vuai defined his role as state-appointed *kadhi* solely in terms of religion, and claimed to defer neither to the state or the community in his practice. However, he was aware of regulations concerning witnesses, and told me that he was able to adhere to the Kadhis Act and to Islamic procedural law. When I asked how he did this, he gave a rather ingenious explanation: because of the act, he could not require the testimony of two female witnesses to equal one male witness. However, if a woman brought in one female witness and her husband brought in one male witness, and he determined that the woman's case was stronger, he could decide in her favor according to Islamic rules by basing his decision on the litigant's testimony instead of that of the witnesses. Therefore, he would not need to call another female witness, and would not violate Islamic or state procedural law. Shaykh Vuai explained that in his ruling, he would write that the witness testimony illuminated the litigant testimony, but was not the basis for the decision, thus satisfying both state and Islamic law. In a 2002 interview, however, this *kadhi* claimed he would only follow Islamic rules of procedure.

In their present-day form, the Zanzibar *kadhi*s' courts are established by a state that modifies Islamic family law by limiting jurisdiction and designating rules of procedure. A *kadhi*'s reasoning and court practice should thus be considered with reference to his political role as a religious authority who is appointed by the state and works in a community. Shaykh Hamid's incorporation of local *kadhi*s, *sheha*s, and elders in the proceedings as witnesses and advisors situates the court as part and parcel of the local community, not simply as an arm of the state. Shaykh Hamid certainly respected the limits of his jurisdiction in religious matters as a state employee, but remained embedded within, and embedded the court in, the community by integrating the expertise of local authority figures into courtroom practice. Indeed, although I frequently heard *sheha*s criticized as political agents, I did not hear this type of criticism leveled against Shaykh Hamid. Shaykh Hamid, who claimed political neutrality, seemed to maintain authority in a community that harbored a strong opposition to the ruling party by deferring to local authority figures regardless of their political leanings; he did not appear to preference the state or local modes of authority in his courtroom, which seems to mirror his joke that a *kadhi* "could be a member of both parties or neither." In this way, we see that the courts can be sites of accommodation for judges who are positioned between the state from which they derive their authority and communities that may be suspicious of the state but amenable to other forms of authority.[15]

CHAPTER SIX

Buying Divorce through Khuluu

A *kadhi* must navigate between his understanding of Islamic law, secular state law, and local practices of marriage and divorce, and in this chapter I explore the way in which Shaykh Hamid understood and utilized one type of Islamic divorce, *khuluu* (Ar. *khul'*) vis-à-vis local cultural norms of marriage and divorce. Scholars have sometimes characterized *khul'* as a divorce by mutual consent (e.g., Esposito 2001), and others have described it as divorce by women's initiative; Tucker refers to *khul'* as a woman's divorce (1998). In the Mkokotoni court, however, we will see that *khuluu* is also enacted by judicial ruling or suggestion, and is thus certainly not always a divorce initiated by women or one of mutual agreement. The same seems to have held true in Zanzibar for quite some time, as Stockreiter has also made a similar point based on her research on Zanzibar Town courts between 1900 and 1963 (2008: 217), when *kadhis* were very willing to offer women the opportunity to buy divorce in *khuluu* (230).

In the Islamic tradition, the permissibility of *khuluu* is most often tied to Qur'an 2:229: "you are not allowed to take away the least of what you have given your wives, unless both of you fear that you would not be able to keep within the limits set by God. If you fear you cannot maintain the bonds fixed by God, there will be no blame on either if the woman redeems herself."[1] Most scholars take this to mean that a woman returns some or all of her dower (*mahari*, Ar. *mahr*) to secure divorce (Esposito 2001). In the colloquial Kiswahili of rural Zanzibar, this is termed "buying a divorce" and although this does happen out of court, the goal of this chapter is to show how and when Shaykh Hamid utilized *khuluu* in court and how the appropriate compensation was determined. As we will see, the *kadhi* did not always utilize *khuluu* simply according to his understanding of Islamic law, but took into consideration who was at fault in marital discord according to local standards of marital behavior and what constituted a fair sum in the local context.[2] The *kadhi* viewed *khuluu* not simply as a divorce by female prerogative, but potentially useful as a compensatory or punitive measure, which provides us an opportunity to gauge how Shaykh Hamid incorporated ideas of equity and fairness into arbitration. At times, he called a *khuluu* settlement a "fine" incumbent upon a wife for her bad behavior in her marriage; the following cases illustrate, correspondingly, how fault and blame were ascribed and assessed in disputes. Sometimes, Shaykh Hamid used ideas of fairness and fault to override what he considered technically lawful in *khuluu*. In such cases, he neither deemed the matter to be outside the bounds of religious law, nor used a general Islamic legal principle to supersede a specific rule; rather, he explained that he handled the matter in accord with community norms.[3]

Marriage and *Mahari*

As elsewhere in the Muslim world, an Islamic marriage in Zanzibar is not considered complete unless the amount of the *mahari* has been written into the wedding contract and at least some of the amount has changed hands.[4] Once all parties have agreed on a marriage, the parents or guardians of the bride determine an appropriate amount for the *mahari* and submit it for approval to the parents of the groom, or sometimes to the groom himself. If the bride has been married before, she will often contract the amount of the *mahari* herself. In rural Unguja, the *mahari* is often paid in full at the time of the marriage. When women opened cases in the Mkokotoni court, the clerks always asked the amount of *mahari* and whether she had received it in entirety. Following their lead, I asked the same questions in my interviews. Only a few women said that their husbands had not paid the entire amount; this was further confirmed by the frequency with which husbands ask their wives to return the entire *mahari* upon the divorce.[5] Also, I knew of very few cases in which women sued for the remainder of the *mahari;* from January 1999 to July 2000, only two women opened cases involving a claim for unpaid *mahari.* Interestingly, this is rather different from what J.N.D. Anderson reported of Zanzibari marriages 50 years ago, when he noted that the *mahari* payment was often deferred, sometimes indefinitely (1950: 72). Scholars working in other contexts have noted that *mahr* is often linked to the husband's power of divorce; this happens most often if the dower is deferred, in which case he will owe it to his wife if he divorces her (Esposito 1982, Moors 1995, Mir-Hosseini 2000). Although deferred dowers appear to be rare in rural marriage contracts, *mahari* is still strategically linked to the husbands' power of divorce, albeit in a slightly different way: women prepared for the possibility that a husband might ask her to return the money in exchange for divorce.

The *mahari* is provided by the groom alone or by his family; the family may also provide *sanduku* to the bride, which is often a gift of clothing and is considered separate from the *mahari*. Several people told me that in the recent past, when elders arranged most marriages for young men, the groom's parents were more likely to provide the *mahari*, especially for first marriages. My gentle neighbor Mzee Farad told me that when he was married for the first time, he had no idea of the *mahari* amount his parents gave. However, in his most recent marriage in the early 1990s, he paid the entire 4,000 shilling *mahari* himself. Today, it is not unusual for men who are quite young to take complete responsibility for the *mahari*. In recent years, a typical *mahari* for a first marriage in the rural areas could range from 10,000 to 150,000 shillings (10–150 USD in 2005).[6] The amount varies widely depending on the wealth of the families involved, whether the bride has been previously married, and her education level. The worth of the family's property, like a house, may also be taken into account. The relationship of the bride to the groom's family is also considered, and if she is a close relation, a smaller *mahari* might be agreed upon. The teacher Mwalimu Adamu told me that he gave only 7,000 shillings when he married for the third time because his wife was the daughter of his maternal uncle. In town, the average amounts are usually far higher. Bi Kala, a recently married woman in her mid-thirties, said the amount could range from 100,000 to one million shillings and noted that one of her friends recently received a *mahari* of 500,000. A male friend from a reasonably well-off family married for the first time in early 2005, and told me he gave 200,000 in *mahari,* but also provided several household items for his bride, including furniture. My "mother" in Zanzibar Town, Bi Serena, who had several married daughters, thought it distasteful to ask for a great deal of money because, "You don't sell you daughter!" In both locales, however, it is somewhat difficult to determine average amounts because people are often reluctant to reveal the exact *mahari* in their own marriages, particularly recent marriages. The topic is considered by some to be a private matter, and I did not push the subject if people hesitated.

I did, however, hear many complaints from men about how difficult it was to marry these days because women expected such a large *mahari*. Always, though, the amount of the *mahari* was discussed in terms of buying power and with a consideration of the value of a shilling in the present and in the past. I discussed this once with Mzee Haji, in his seventies, and Mzee Issa, about 50. They were a jovial duo who often dropped by our house to chat when the other men in the community were at the mosque for the evening prayer. They made quite a pair: the short Mzee Haji retained a charmingly boyish face even in his old age, and Mzee Issa was tall and robustly built; he was nearly twice Mzee Haji's size. Mwanahawa and I enjoyed their visits. Mzee Haji liked to tease her, and insisted that I take photos of them posing together so he could pretend she was his wife, which made Mzee Issa chuckle heartily. He was a joyful man with a big belly and a big smile who was much loved in the community. When he died suddenly in 2002, his death was mourned for a long time.

I knew them well, and we talked about my research on many occasions. One evening, I interviewed them together about marriage, and both told me that their parents provided the *mahari* for their first marriages. Mzee Haji said his was only 120 shillings but clarified that "these were not like the shillings of today." He added, "Now, a *mahari* is typically 100,000 shillings." Mzee Issa shook his head in agreement at the high rate and added, "Or even five *laki* (500,000). My *mahari* was 280 shillings."

> "But in the past, were you able to buy a lot of things with 100 shillings?" I asked.
>
> "Oh yes," answered Mzee Haji. "It had more buying power than 100,000 shillings today!" Mzee Issa added, "Back then, with 300 shillings you could buy everything you needed for the house—the goods, the bed, the food. And everything you needed concerning clothing."

Both today and in the past, it is quite common that a bride will not receive the *mahari* herself, and several people explained that a woman's parents would collect it and give only a small portion or nothing at all to the bride.[7] In interviews, women consistently told me that they were not given their *mahari* at the first marriage; nearly all said their parents took it, although a few received it directly on subsequent marriages. It should be noted, however, that parents may use the *mahari* to buy a bride the goods to take to the marital home, such as kitchen and washing paraphernalia, a bed, and bed linens; many women said that they knew that their parents used some of the *mahari* to buy these *vyombo*.[8] Bi Jabu, a woman of about 60 who had been married many times, did not know the amount of the *mahari* in her first marriage but her parents gave her 400 shillings to buy marriage goods; she did not know if they received more than that amount. The expressive Bi Pili, in her seventies, said, "I didn't know [the amount of the *mahari*] because I wasn't told. In those days you didn't know, the *wazee* just took it! They just took [the money] in those days!" Kaeni told me that her elders took some of the *mahari* to buy her goods:

> It was only 700 shillings. In those days, there wasn't a lot of money. My elders gave me 500 shillings. Your parents got the marriage gift—they told me that the *mahari* was there but that they would use it to find a bed.

She went on to explain that one of her elders had died before he gave her all of the money, but that she forgave him the debt because she felt that he had meant to do the right thing, and thought that "it was truly his goal to find something for me."

Bi Asia, in her early thirties, told me that her parents took the *mahari* for the first marriage, but used it to buy things for the house. She explained that the husband builds the house;

> But then the wife brings everything else—the bed and all the things for cooking. The husband doesn't buy anything—he only gives the *mahari* then the wife buys the things. He also buys the clothes that you'll wear at the house—dresses, several *kanga*s, slips, and underwear.

When I asked her if it was necessary that a bride must buy more goods after the wedding, she replied,

> If you have many things, then you don't have to buy any again. However, if he breaks them then you'll have to tell your husband to give you some money to buy the things again. And then he'll give the money and you'll buy them again.[9]

As we saw in chapter three, elders often have less influence in second or third marriages, and women often negotiate the *mahari* amount themselves. One woman called Mwashamba explained that she chose her second husband herself and that accordingly she took the *mahari* of 47,000 shillings. She explained that she contracted the entire marriage without help from her elders, and when she told her fiancé how much she wanted for the *mahari*, he approved: "He asked me—what amount of *mahari* do you want? I just told him and he agreed." She noted, however, that he still owed her 17,000 shillings of that amount.

A woman in her thirties called Halima, whom I knew as "the doctor" because she worked at the health clinic, shed some light on what a woman might consider when negotiating her own *mahari*. I had scheduled an interview with her elderly aunt one afternoon, and as we sat and talked, Halima joined in. We were discussing *mahari,* and Halima laughed as she explained that elders always take the *mahari* because "they have their stomachs to fill!" When I asked them how the amount of the *mahari* was determined, Halima was one of the few women who noted that it depends on the resources of the man. Later, she asked us rhetorically if a woman and her elders would refuse a man simply because he was not rich and could not provide a large *mahari*: she implied that this would be foolish, and added "*kila mtu na uwezo wake*" (every person has his own ability). Halima also emphasized that sometimes women did not want a large *mahari*. Addressing my curiosity, she explained that some women avoided a sizeable *mahari* because "if he divorces you, he'll say he wants his money back and then you won't be able to give it to him...he'll want his money back and you'll have to give him that very amount [of *mahari*]!" Although many women talked to me about the paying husbands for divorce, this was one of the few instances when someone specifically connected it to the amount of *mahari*, noting that women may consider the unfortunate prospect of "buying" a divorce when negotiating their *mahari*.

Conversations in 2002 with the rambunctious Rehema (then about 18) and her petite and impish friend Mwanaharusi (19) illustrate that younger women's expectations about *mahari* may not match the reality. The two girls were very interested in discussing marriage because rumor held that Mwanaharusi had recently received a proposal and that her mother was planning a wedding. Mwanaharusi claimed ignorance, but Rehema was eager to speculate about the imminent nuptials. I asked how the negotiations were coming along; Rehema told me that the *mahari* had been agreed upon, but that they did not know the amount. Mwanaharusi confirmed that she had not received anything, to which Rehema replied, "Humph! Your grandfather should not take his hand [i.e., make an agreement with the groom] until you are given the money!"

When I asked why, Rehema shouted as if I was both dim-witted and hard of hearing, "*Haki* (it's your right)! You must get your *mahari* as a woman!"

I said that I thought sometimes the elders took it, and she explained that "The elders take it, but they give to the children. It's your *haki*. If someone gets a proposal, she might get 30 shillings or 300,000, but as a woman you get it yourself."

I asked how they knew about this right and Rehema, frustrated with my ignorance, repeated, "It's just your rights!" Mwanaharusi was more patient with me and explained, "You hear it on the radio...at *chuo* (Qur'an school)...It's just *haki*. Your father also tells you. If someone proposes to you, then you get the whole *mahari* yourself. Then you go and get whatever you want."

Rehema was delighted at this prospect, and enthusiastically informed me that indeed a bride could buy whatever she wants, "Gold, rings, earrings, bangles...you can get whatever you want! You're given all the money!"

The discussion continued, and I noted the fervor with which the girls told me about their rights and how they had learned of them. I was especially interested because the girls were not outstanding students in either public school or Qur'an school, and I doubted they had poured over many books. Rehema, though clearly bright, never really learned to read although she went to the local school for many years. They both had left school at about age 17.

Three years later, in 2005, I was sorry to learn that the reality of Mwanharusi's marriage had not met the girls' sunny expectations. She had indeed been married shortly after my last visit to a young man who lived in village in another part of the island. She was now 22, and though she retained her girlish energy and sense of humor, she had become a grown woman with a woman's concerns; she and the as yet unmarried Rehema rarely saw each other (Rehema married in 2008). Mwanaharusi had returned home to her mother to give birth to her second child, and she was still recuperating there when I visited. Her baby was a handsome boy, and though she was pleased to have another child she was also concerned: the baby had not yet been circumcised because they did not have money to pay for it. This anxiety led her to tell me about her greater financial worries, which centered on the *mahari*. There had been no carefree days of spending the *mahari* on gold earrings and necklaces, like Rehema had imagined. Her father, who had divorced her mother, Bi Safa, many years ago, had arranged her marriage. I asked her if she had received her *mahari*, and she told me that she had received a portion and was waiting for the balance. Of a total of 70,000 shillings, she had been given 8000 shillings on her wedding day. The problem was not that her husband had not brought the money, but rather that her elders in her father's village were keeping it for themselves. She had been asking for the money from the day she got married, but was still waiting; "They are treating me badly!"

When I asked her about the relevant *sheria,* she told me that according to religious law, she should have received all of the money. I asked her if the elders had bought her household goods with the money, but she replied, "They didn't buy me a bed; they didn't buy me clothes; they didn't even buy me dishes; this was all the responsibility of my mother." I knew her mother well, and understood how difficult the situation was. We were neighbors, and often helped each other out in small ways. I would help with cash when Bi Safa had expenses she could not meet (like the new grandson's circumcision), and she would bring me vegetables, oil, or fresh fish from her husband, who was the gentle Mzee Farad. However, Mzee Farad was an increasingly sickly man and was often unable to farm or fish. Thus, they remained among the poorest families in the village even though Safa, who was much younger than her husband, worked her fingers to the bone farming and pressing coconut oil for sale. Although there was certainly no extra money for wedding goods, Mwanaharusi told me that her mother managed to buy her some dishes and buckets for washing even though she did not have the *mahari* to draw on. Her paternal grandmother also gave her a few small items, but the major essentials like a bed and mosquito net never materialized. Despite these difficulties and the new baby, Mwanaharusi told me she was determined to demand her rights and was hopeful that someday her elders would give her the *mahari.* Bi Safa also continued to persevere. Although her life was difficult, she had nothing but the kindest words about her sweet-tempered second husband, and she missed him greatly when he died in 2008.

Zanzibari religious scholars and authorities often expressed their disapproval of *mahari* practice. One *kadhi* told me that ideally, the *mahari* should be "like the ring in your [American] marriages: it is a token of love." He lamented that many Zanzibari parents did not understand this, and took the *mahari* for themselves. Sadly, he said, "The elders think that anything that belongs to their children actually belongs to them." Hirsch has noted that Kenyan *kadhi*s similarly insist that the *mahari* belongs to the bride (2008). Furthermore, many people in Zanzibar—religious experts and lay people—described the *mahari* negotiation as something in which many relatives request or demand a portion or require the groom to give specific gifts to a girl's family. Most were critical of this and considered similar practices as a violation of "proper" *mahari* according to religious law. For example, Shaykh Hamid explained one case in which a young woman's elders coerced a young man by asking for expensive gifts for themselves in addition to the *mahari*. He said that this was inappropriate and was akin to bribing the young man.

Buying a Divorce in Court

As we have seen, many of Shaykh Hamid's rulings reference *khuluu* divorce, as in the *masharti* that specified that if a woman breaks the terms she must buy her divorce. Shaykh Hamid also ruled for judicial *khuluu,* and such rulings normally took into account the amount of the *mahari* in determining the appropriate amount of *khuluu*. These cases are particularly interesting because although Shaykh Hamid often asserted that a man had no right to ask for more than the *mahari* in a *khuluu* divorce, he supported the practice and even encouraged it under certain circumstances. The first time we discussed the matter, he said he permitted it because the disputing parties had come to an agreement outside of the courtroom that the woman would pay more than her *mahari* for *khuluu*. Hence, even though it was technically unlawful, he allowed it if the amount was negotiated out of the courtroom by the litigants and their elders. After this conversation, I surmised that the *khuluu* amount could be negotiated in any case in which a stubborn man demanded more to divorce his wife than what he had paid her in *mahari*. As time passed, however, I noticed that it was more complex. It was only in certain cases and under specific circumstances that Shaykh Hamid allowed the amount to be negotiated. These were those cases in which he determined that the wife was more to blame for the marital strife than her husband. For Shaykh Hamid, then, *khuluu* was associated with who is at fault in the marital discord as much as it is with who initiated the divorce.[10]

Although the *kadhi* could not influence divorces that took place out of court, he did make a point of making sure that alleged divorces were proper *khuluu* instead of instances of "writing for money." Let us briefly revisit Machano's case from chapter two. In this case, Shaykh Hamid eventually ruled that Aisha had been validly divorced from Machano "according to the laws of Islam" through *khuluu,* which he had determined through questioning the litigants and witnesses.

He determined that the divorce was appropriate *khuluu* and not unlaw-ful writing for money because although Machano had tried to return to Aisha, her refusal indicated her desire to be divorced from him. This was an appropriate instance of *khuluu* because she desired the divorce.

Mosa's case, also from chapter two, showed Shaykh Hamid's reason-ing process in determining when a woman could buy divorce. Recall that Mosa, under the guidance of her clever daughter, strategically described marital problems in an attempt to avoid paying for a *khu-luu* divorce. Their strategy did not work, however, because the *kadhi* determined that Mosa was at significant fault in the marital problems. Mosa eventually bought a divorce from her husband because she had violated the terms of reconciliation. When her husband Juma asked for 70,000 shillings Shaykh Hamid explained that he was breaking the law by asking for this amount since it was 10 times her *masharti* of 7000 shillings. However, he allowed them to negotiate the amount if they left the courtroom for the negotiation process. They eventually settled on 25,000 shillings, an amount much greater than the *mahari*. Shaykh Hamid explained that the negotiation was permissible because Mosa was at greater fault in the marital discord and therefore should pay Juma more than she received in *masharti* out of fairness: in addition to causing marital strife by allowing her children to berate Juma, as was established by witnesses, she failed to return to him as ordered in the *masharti*. Furthermore, she failed to follow court-ordered procedure that to see the *sheha* if she had more problems with Juma. The written ruling, however, noted that this was indeed the amount of the *mahari*: "...the court wants the plaintiff to pay the amount with which she was married [*mahari*]; 25,000 shillings."

In Zaynab and Rashidi's case from chapter three, Shaykh Hamid also ruled for reconciliation. Like Mosa, Zaynab never returned to Rashidi, and clearly had no intention to do so. Thus, because she vio-lated the *masharti*, she had to buy her divorce. Rashidi requested 80,000 shillings, but she protested that this was far too much. After some dis-cussion, it was clear that the two did not agree on the amount of the original *mahari*, and so Shaykh Hamid set an amount somewhere in the middle of what each claimed. In the write-up, however, Shaykh Hamid indicated that the amount was equivalent to the *mahari*. It was not noted as a "fine," as he sometimes did in other rulings. This was because Rashidi's only complaint was that she would not come back to his home and Shaykh Hamid did not think she had caused many prob-lems in the marriage. Furthermore, although there was no evidence to support Zaynab's claim that she had been divorced, she seemed to truly believe that she was, had not received maintenance, and claimed to have been verbally abused by her husband. Thus, in Shaykh Hamid's eyes, she was at very little fault in the marriage problems, and was not expected to pay *khuluu* as a fine for her transgressions.

Another Wife for Samir: *Siri v. Samir*

This case, opened by a young woman called Siri, involves a negotiation of the amount paid in *khuluu*. When Siri came to court, her father accompanied her and explained that she wanted to open a case against her husband. Although her father began the discussion, Siri was open about her problems, and was not shy to tell the *kadhi* that she no longer got along with her husband. She said the house was in a poor state of repair, and that she thought her husband did not want her anymore, because a man who cared for his wife would certainly make repairs to the house. So, she said, she left to return to her father. Her father jumped in at this point to make clear that they had waited a long time for her husband to improve, but that she still had no clothes and no decent house. The *kadhi* explained that they could not summon her husband until a case was opened, so he sent her to the clerks to prepare a claim. The document was prepared as a divorce request, and noted that Siri's husband threatened her; I am not sure why the maintenance was not included—I was listening to another hearing while the clerks prepared this claim and so did not hear their discussion with Siri.

Madai

1. The plaintiff is a woman, aged 29.
2. That the defendant is a man, aged 36.
3. The plaintiff claims that she and the defendant are wife and husband and they have been married for seven years and have one child.
4. The plaintiff wants a divorce; she does not want the defendant, her husband, because they do not get along.
5. That basis of this claim is that the defendant said violent words to the plaintiff; he told her that he "would make her lame and end her life."
6. This claim originates from D—.
7. The plaintiff asks the court to rule that the defendant follow what is written below:
   a. The defendant divorces the plaintiff.

b. The defendant must pay all court fees.
c. The defendant must follow any other orders issues in the agreement of the plaintiff.

When Samir came to court, he said that he had no desire to divorce her. Furthermore, he claimed that he had neither verbally abused her nor failed to support her.

Majibu ya Madai

1. The defendant agrees with the first three points of the plaintiff's explanation.
2. The defendant does not agree with point 4 of the plaintiff's claim and explains that he, the defendant, does not want to divorce the plaintiff; he has not refused the plaintiff, but notes that she has some problems with the marriage.
3. The defendant does not agree with the explanation in point 5 of the plaintiff's claim and explains that he, the defendant, had not intended to "make her [the plaintiff] lame." The plaintiff and the defendant went on a trip to Tanzania for one year and seven months; the defendant never did anything to disrespect his wife.
4. This claim originates from D—.
5. The defendant asks the court to rule that the plaintiff follow what is written below:
   a. The court should dismiss the plaintiff's demand for divorce because the defendant has not refused his wife, and does not want to divorce her.
   b. The court fees paid by the plaintiff.
   c. Any other orders issued in the agreement of the defendant.

On their first day in court together, Shaykh Hamid gave his familiar explanation that the court was like a hospital, and he would solve the problem the way a hospital would cure an illness. When he invited Siri to speak, she explained that three years earlier Samir had "written for money" asking for 20,000 shillings to divorce her. They had gone to the *sheha* together, where he again asked for money, this time increasing the amount to 40,000 shillings. She did not give him the money either time. Samir, a mild-tempered young man who rarely showed any emotion during the proceedings, confirmed that they had gone to the *sheha* about their problems, and he said that he had told Siri that he would divorce her if she paid him 20,000 shillings. Siri had left him to live with her parents before going to the *sheha,* and he waited 11 days but heard nothing from her and received no money. He explained that he told the *sheha* and Siri's mother and father that he wanted her back, but she refused.

After hearing their explanations, Shaykh Hamid said that even though they had not been living together for some time, there was not yet any justification for a divorce on the grounds that Samir had failed in his marital duties. He encouraged them both to bring witnesses to court, and they did so a few days later. None of the witnesses established that Samir had failed in his marital duties, and so the *kadhi* tried to determine why Siri had left Samir because it did not seem he had done anything wrong. At this, however, she turned inward. She looked at the floor, played with her hands, and said she was "too shy" to talk about it any further. Seemingly at a loss, the *kadhi* eventually encouraged the couple to reconcile. Siri seemed unhappy, and when the *kadhi* asked her slowly and clearly if she wanted her husband, she was initially reticent but shyly answered "*Simtaki*" (I don't want him). When Shaykh Hamid asked if she was willing to "buy" her divorce, she readily agreed. Samir was very calm throughout, and he was so quiet that the *kadhi* asked him if he had heard that his wife wanted a divorce. He had, and simply said that she should buy the divorce for an amount that would allow him to remarry, "I should get my money so I can marry another wife."[11] The *kadhi* asked Siri if she heard him, and agreed that Samir needed enough to remarry, "He can't sleep by himself, and since you won't go to him, who will he sleep with? A man wants a wife."

Shaykh Hamid asked Samir how much he wanted, and he answered that he wanted 100,000 or 200,000 shillings. This was a large sum for a *mahari* in the area, and after protests from both Siri and the *kadhi,* Samir conceded that 70,000 shillings would allow him to remarry. Siri, who was rather sullen at this point, protested that this was still far too much. The *kadhi* then asked them about the *mahari,* and they both said that it was 15,000 shillings. Shaykh Hamid made a point of explaining that Samir had no lawful right to ask for more money than he paid in *mahari.* (I wondered if Shaykh Hamid had regretted asking Samir how much he wanted before asking the amount of the *mahari;* this may have been a mistake.)

Shaykh Hamid asked Siri how much she would give Samir "to get a new wife." She did not answer right away, so Shaykh Hamid asked Samir to leave the room with him to talk privately. After a few minutes, he sent Samir back inside and called Siri out to discuss the matter. They eventually agreed on a sum of 40,000 shillings in these negotiations that were supervised by Shaykh Hamid but conducted outside the walls of the courtroom. Shaykh Hamid decided that Siri would have two months to pay it in full. Upon this announcement, Samir's satisfied smile melted. He was not pleased at the thought of waiting so long and made it known. The *kadhi* answered his complaint by telling him that he could not force Siri to pay immediately. To Siri, he explained that the divorce would not be considered final until she brought the money to the court. He then announced to the room, "If she brings it tomorrow, then *sheria* will be fulfilled tomorrow." The written ruling required Siri to pay 40,000 shillings and to return to her husband until she did so.

> In the court of the *kadhi* of Mkokotoni, Unguja, the plaintiff and the defendant are here in the court. After listening to their claims and explaining various things about the law, the plaintiff says, "I don't want my husband. And if he divorces me free, fine then I will buy his divorce lawfully." And the defendant says, "If she wants me to divorce her, fine." The plaintiff Siri should buy her divorce and she agreed to buy her divorce from her husband for 40,000 shillings. And the defendant has agreed to have the divorce bought from him for 40,000. Therefore, the court issues the judgment according to the law that the plaintiff buys her divorce for 40,000 and she has two months to pay it beginning February 7, 2000 until April 1, 2000. She will pay this money here in the court in this period, and she is not divorced by her husband until the bail amount has been paid to the court and received by her husband. It will be the responsibility of the husband to collect this money. No one else can do so. This is finished, and indeed, on the day she brings the money she will be divorced from her husband.

> The court has decided that the plaintiff, Siri, must follow her husband home and stay with him. This begins today February 7, 2000 until April 1, 2000. She must bring the money here to the court in the amount of 40,000 to buy her divorce from her husband. She has agreed.

After he read the decision aloud, Shaykh Hamid explained again that *khuluu* should never exceed the amount of the *mahari*. However, in this case the amount was greater than the 15,000 shilling *mahari* because Siri left and refused to return to her husband even though she gave no evidence of his failure to maintain her or his abuse. Shaykh Hamid reasoned that the mistake was hers, and that she must therefore pay a "fine" in the divorce, indicating that the *khuluu* payment not only compensated Samir for the loss of a wife but also punished Siri for her failings in the marriage, and he referred to the amount as "bail" in the ruling. As in Mosa's case, Shaykh Hamid reasoned aloud that the wife was at fault in the marital discord, and as a result, he ordered each to pay for the *khuluu* divorce with an amount greater than her *mahari*. In both, the decision included the amount to be paid in *khuluu,* and was written as an agreement between the parties based on the woman's refusal to return to her husband despite his entreaties; in neither ruling did he note that it exceeded the amount of *mahari*.

When the *kadhi* was finished reading, Siri left the court with an odd look on her face. I felt a bit unsettled, and wondered if there were problems in the marriage that we never knew about. The *kadhi* continued talking with Samir for some time, once again telling him that buying a divorce should not exceed the amount of the *mahari*. Although he told Siri that she must return to Samir until she had paid the 40,000 shillings, she never went back to him. She paid the sum about two months later.

Sheria za Kikwetu (Local Law): *Hamdu v. Rukia*

The next case involved a highly unusual variation on *khuluu*. The plaintiff was a very small, rather fussy man in his fifties named Mzee Hamdu. On his first day in court, he brought a letter from his *sheha*, which the clerk read aloud. I was listened eagerly, because although litigants were supposed to have letters from *shehas* before opening cases, I rarely got to see them. I was a bit disappointed when the letter simply read: "This person has a problem with his wife, so please listen to him." Hamdu, however, elaborated. He told the *kadhi* and clerks that his wife, Rukia, had left him. He launched into a very long explanation of how some months earlier she went to visit an ill family member in another village but had not returned to him and their children. The *kadhi* told him to open a case, which he was eager to do. The *madai* was written as a simple request for the return of an absent wife.

Madai

1. That the plaintiff is an adult man aged 50, Mtumbatu, from
C—.

2. The defendant is an adult woman aged 36, Mtumbatu, from D—.
3. The plaintiff and defendant are wife and husband, and they have been married for 26 years. They have nine children, one of whom died.[12]
4. The plaintiff claims that his wife left him, and has not been at his home for six months; he does not know where she is.
5. The plaintiff claims that three months ago he saw his wife at M— village at the house of the late M.F.
6. The plaintiff claims that he no longer has the ability to persevere in waiting for her return; the plaintiff went to the defendant's elders but she has still not returned to him.
7. The basis of this claim is that the defendant has left the plaintiff, her husband, without permission.
8. The plaintiff asks the court to listen to his claims originating from C— in the Northern A district of Unguja.
9. The plaintiff begs the court of the *kadhi* to rule that the defendant follow what is written below:
   a. The defendant returns to the plaintiff immediately.
   b. The defendant pays the court fees if she will not return to the plaintiff.
   c. Any other orders issues in the agreement of the plaintiff.

Bi Rukia came to court a few days later. She was a plump and attractive middle-aged woman with a calm, intelligent expression in her large eyes; she probably weighed twice what her diminutive husband did. She was wearing a striped dress and a *buibui*. It was clear from her first moments in court that she had no interest in her husband. She told the clerks that she had been absent from their home for a while, but explained that this was because her *mlezi*, the woman who raised her, had become ill and died. Her *majibu* stated that it was not true that she left Hamdu without permission as he claimed. Rather, she left to be at the bedside of her dying *mlezi,* which she had arranged with his approval. The document also stated that after the death, there were some legal problems concerning the ownership of the deceased woman's house, and that she had received permission from her husband to stay there until the problems were solved.

Majibu ya Madai

1. The defendant agrees with the first three points of the plaintiff's claim.

2. The defendant does not agree with the explanation in points 4 and 5 and explains that it is not true that she, the defendant, unlawfully left her husband the plaintiff; she left according to their agreement that she go to M— to care for her ill *mlezi*. There were no problems concerning this between the plaintiff and the defendant.

3. Concerning points 7 and 8 of the plaintiff's claim, the defendant explains that after the death of her *mlezi*, there were some problems with the deceased's home because the government wanted someone to live inside it until the problems concerning the house were solved. The plaintiff agreed with the defendant and gave her permission to stay at the house.

4. The defendant claims that she is tired of the lies of the plaintiff, and she begs the court to rule that the plaintiff follow what is written below:

   a. The plaintiff divorces his wife one time.

   b. The plaintiff pays court fees.

   c. Any other orders issued in agreement with the defendant.

It was an unusually cool and rainy day when they came in together to see the *kadhi*. Hamdu told the *kadhi* that his wife had left him, and with great animation, explained that she had wanted a divorce for some time, but that he did not want to divorce her. Throughout his testimony, Rukia avoided looking at him. She seemed exasperated and often rolled her eyes at his comments. Though they were sitting on chairs side by side in front of the *kadhi's* desk, she turned her body away from him and was essentially sitting on the edge of the chair in order to put as much distance between them as possible. When it was her turn to speak, she said that she had gone to care for her ill relative after taking leave of Hamdu properly. She complained that he had not come to see her even once while she was gone, and agreed that she had wanted a divorce for a long time—she had even offered buy it from him with her *mahari*. Hamdu never accepted her offer, however, and had tried to forbid her from going to the *kadhi* to seek a divorce.[13]

It was evident that Rukia did not want her husband. I thought her loathing of him was obvious to us all, and I was thus somewhat surprised when Shaykh Hamid optimistically expressed his hopes for their potential reconciliation and his decision to issue *masharti* that required Rukia to return to Hamdu. The *kadhi* explained that he would give them a *muda* (specific time period) in which they must try to carry out the *masharti*. As usual, he took great care to explain that if the reconciliation did not work out, then they would be divorced, and noted that if Rukia broke the terms then she must buy her divorce (which is what she had been trying to do for some time!). Hamdu looked well satisfied with this and nodded eagerly at everything the *kadhi* said. Rukia was displeased, however, and explained one more time that she did not want her husband. When Shaykh Hamid cheerfully tried to convince her to persevere, she looked at me and smirked. Undaunted, the *kadhi* continued. "Yes!" he said, "You *say* you do not want him, but persevere a little bit. Persevere a little bit and then perhaps you'll get a divorce!" The mention of divorce lifted her spirits a bit, and she waved at me cheerfully as she left.

They were back a few days later for the written *masharti*. Rukia was wearing the same dress, but Hamdu's jaunty pair of striped trousers and pink shirt matched his jolly mood. The *kadhi* explained the terms of the ruling, specifically telling Rukia that if she did not go back to her husband, she would be divorced. After this, he donned a pair of sunglasses and left, complaining of a headache and leaving Bwana Fumu to read the *masharti* aloud. I thought the whole production of reading the document and getting signatures a bit futile because it seemed clear that Rukia did not intend to fulfill any part of it.

Hukumu

After listening to the plaintiff and the defendant together and with their witnesses of both sides, the court decides that both of them have made mistakes and therefore the court gives them terms.

Terms for the Plaintiff

The plaintiff must know that he indeed is the leader of the wife and their children and he must only speak the truth and speak kindly with his wife. He is ordered to get along well with his wife by fixing their house and supporting the children and his wife according to the law. He is expected to fulfill these terms habitually, and if he does not fulfill them, he will have violated the law and his wife will be able to break the marriage and be divorced.

Terms for the Defendant

The defendant is ordered to follow her husband back to his home and to fulfill all of the matters of the marriage laws of husband and wife. Because your husband has not yet divorced you, the defendant, you are expected to live well with him without irritating him or rebuking him and you should consult with him on matters of the home. Also, if you have a journey you must take proper leave of your husband. You are expected to follow these terms because if you do not you will have violated the law. If you breach the law, you will need to buy your divorce according to the law of *khuluu*.

And if it so happens that either party makes a mistake, then the plaintiff or defendant must go to the *sheha,* who will be expected to write a report, and the court will fulfill the orders of the law. This indeed is the judgment.

To no one's surprise except maybe Hamdu's, they were back in court one week later. Crossly, Hamdu told the *kadhi* that Rukia had not returned to him as ordered in the *masharti*. Rukia readily agreed that she had not gone back, and immediately showed the *kadhi* the 30,000 shillings that she brought to "buy" her divorce. She had never intended to return to her husband and had been fully prepared for *khuluu* as outlined in the *masharti*.

Seemingly unconcerned at their failure to reconcile, Shaykh Hamid immediately moved to prepare for *khuluu*. However, Hamdu angrily refused to take her money and made a huge fuss about it. The *kadhi* eventually tired of his protestations, and devised an unusual method to rectify the situation: he suggested that Rukia return to her husband for another prescribed period of time and receive a divorce at the end. In return, Hamdu would relinquish his rights to any financial compensation for the divorce. Hamdu liked the plan and his mood brightened instantly. He said that the time Rukia spent with him would be the equivalent of his payment for the divorce. Even Rukia agreed, and Shaykh Hamid suggested a year or six months for the time period. Rukia balked, however, and said she refused to go for such a lengthy period. Eventually the couple settled on only one month.

Before they left, Shaykh Hamid explained the terms of the situation again: if Rukia went back to her husband for the specified period, she would be divorced for "free" at the end of it. He said that this was similar to *khuluu,* but without the exchange of money. After they left, I wondered aloud why Hamdu had refused to take the money for *khuluu,* even when she had it there in her hand. I said that perhaps Hamdu had thought Rukia would decide that she loved him after all if she spent more time with him. Shaykh Hamid laughed, and said he thought the same thing.

We saw them again exactly one month later. Rukia had stayed with Hamdu for the entire month, but still wanted her divorce. Hamdu, however, complained loudly that the situation was "unfair" and said that he wanted his money for *khuluu* since she still wanted a divorce. Shaykh Hamid had lost his patience with the little man, and reminded him that he had agreed to the arrangement: "She doesn't want you! I ordered her to give you money to buy the divorce but you refused it!" Shaykh Hamid would not to listen to any more of the man's protests, and the clerks prepared Rukia's divorce papers.

In this case, Shaykh Hamid developed a creative strategy in lieu of typical *khuluu*. Although at first glance this may look like his attempt to reconcile the couple one more time, it was a pragmatic solution to a difficult case with two stubborn litigants. Even though it appears that Shaykh Hamid knew that Hamdu thought the one-month trial period would be a chance to keep his wife, his order of this arrangement was a clever attempt to appease the troublesome man and give Rukia the divorce. When we talked about the case, he told me that this was a lawful solution to the problem. However, he did not explain it as Islamic or state law, but rather as *sheria za kikwetu*, the local law.

The Elders Are Also to Blame: *Ame v. Zuwena*

The final case illustrates the importance Shaykh Hamid places on assessing faults in determining how much money should be returned in a *khuluu* divorce. Here, the wife's parents' actions figure into Shaykh Hamid's reasoning about the sum involved in *khuluu*. A man in his thirties, Ame, opened a case against his wife, Zuwena, who was a few years younger. Ame claimed that Zuwena left him for no reason, and said that he wanted her back but he blamed her parents for detaining her. Zuwena had been living at her parents' home for several months, and Ame said that he tried to continue to maintain her for the duration, but her parents told him not to come to see her anymore. He asked the court to order his wife to return to him. When I interviewed him, he said that he had waited for several months in hopes that they would change their minds, but they had not.

Madai

1. The plaintiff is a man, aged 35, from F—.
2. The defendant is a woman, aged 25, from K—.

3. The plaintiff and defendant are husband and wife, and they have been married for the past three years. They have three children; two have died and one is living.
4. The plaintiff claims that his wife, who is at her parents' home, has been prevented from returning to him for the past seven months. Now, she does not want to return.
5. The plaintiff has tried to get her to return to him, but he has not succeeded and she has not returned.
6. The basis for this claim is that the defendant is prevented from returning to her husband, at her parents' home and she does not want to return to her husband in the village of F—.
7. This claim originates from Northern A district of Unguja.
8. The plaintiff begs the court to rule the following:
   a. The court orders the defendant to return to the plaintiff immediately if he is her husband.
   b. The defendant must pay the court fees if she does not return to the plaintiff.
   c. Other claims that are in the agreement of the plaintiff .

Zuwena was a stout, shy young woman, who was always with at least one of her parents; she seemed to rely on them a great deal. When they came to court, they stated that she wanted to buy a divorce with her *mahari* of 10,000 shillings.

Majibu ya Madai

1. The defendant agrees with points 1–3 of the *madai,* with the exception that they have been married for seven years, not three.
2. Concerning points 4–5, the defendant answers that this is true, but that there are indeed problems that prevented her from returning.
3. Concerning point 7 of the *madai,* the defendant answers that the claim originates from the village of B—, not K—, as the plaintiff said.
4. Concerning points 7a, b, and c of the *madai,* the defendant does not agree with the plaintiff's demands and responds that the plaintiff is the one who has caused problems in the marriage because he does not support her, the defendant, as he should.
5. The defendant begs the court to rule the following:
   a. The plaintiff takes his *mahari* of 10,000 shillings to give his wife a divorce.

    b. The plaintiff should pay the court fees.

    c. Any other claims in the agreement of the defendant.

In their hearing, Ame simply explained that Zuwena's parents were detaining her. When it was her turn, Zuwena spoke slowly and needed urging and encouragement from her parents. She said that she had been pregnant, and had gone home to give birth about five months before the baby was due. Soon after the child was born, Ame came to take her back home but her parents did not want her to return to him. She suggested that this was because they thought he had not supported her properly when she was living with him. He came again, but she did not go back to him because by then "it was as if she did not have a husband." Shaykh Hamid asked her if she wanted to remain married, and she said no.

One week later, they came back with witnesses. Ame brought his father, who explained that to his knowledge, the couple had had no previous problems. He knew that Zuwena went home and that Ame wanted her back, and he blamed her parents because he thought they were detaining her. Zuwena's witness, her mother, testified that Ame was not supporting Zuwena, and included much detail to evidence his failure to do so. When she finished, Shaykh Hamid said that they should bring a male witness and the *sheha* (this was one of the few times I heard him specify men as witnesses).

Shortly after, they came back with their fathers, and Shaykh Hamid explained at length about the important role of elders in the lives of their children. The witnesses did not bring much that was new: Zuwena's father testified that Ame was not supporting her properly and Ame's father (testifying again) said that he did not understand why the couple was having problems since they had lived together peacefully for many years. Zuwena's *sheha* was also present, but he did not testify. Rather, he assisted the *kadhi* by questioning the litigants. When he asked Zuwena's father if it was true that they had detained Zuwena, her father said that it was true, but explained that they had done so because Ame did not support her. Shaykh Hamid told them that he would decide the case the next week. Before leaving, Zuwena went to talk to the *kadhi,* who told her she should return to her husband. She said she would not go, and Shaykh Hamid joked that in that case she would have to pay several hundred thousand shillings for the divorce.

As in many other cases, Shaykh Hamid issued *masharti,* which ordered Zuwena to return to Ame, and ordered Ame to support her properly. He told the litigants that Zuwena did not get a divorce because she had caused problems in the marriage: She left Ame without telling him where she was going, and stayed away for a very long time without returning. However, Ame had also erred in the marriage because he did not support Zuwena as well as he should have.

Hukumu

After listening to the plaintiff and the defendant together with their witnesses, the court has investigated their claims. Both the plaintiff and the defendant have made mistakes, and the court issues the judgment through terms for the plaintiff and the defendant.

Terms for the Plaintiff

1. The plaintiff must fulfill the laws of marriage by giving his wife sufficient food, he must let her live in an adequate house with a kitchen and bedroom... and it must be near other people who can help his wife with her problems.
2. In addition, he must give her three *kanga*s, two dresses, two slips, one headscarf, and two pairs of panties. He must also give her money for normal incidentals: soap, hair oil, and shoes.
3. You must live well with her and not use harshness or foul language and when you leave you must give enough food to last *until you return* [*kadhi*'s emphasis] from your journey.
4. You must get along with your wife and you are expected to follow these terms; if you do not it will be a violation of the law.
5. If you disrespect your wife, she has the right to dissolve her marriage through *fasikhi* and if she does so, she will be divorced without any payment to you and she will no longer be your wife.

Terms for the Defendant

1. Defendant, you are expected to go with your husband to his home and fulfill his laws of marriage and live well with him; do not use foul language and harshness and get along well with him in matters of the house and serve him well if he asks you to.
2. You are expected to continue to follow these terms, and if you do not it will be a breach of the laws of marriage. And if the plaintiff and the defendant do not fulfill these terms, they must go to the *sheha* who will write a report and bring it to the court.

Some weeks later, Zuwena and her mother came back to court to ask for a more complete explanation of the *masharti*. This trip seemed instigated by the mother as an attempt to argue over the *masharti* rather than to seek clarification. Shaykh Hamid was frustrated with them, and said simply that their case was finished. If Zuwena had any further complaints she must take the matter to the *sheha*. He added that if Ame had not fulfilled his *masharti*, then she would get a *fasikhi* divorce. They left, but the entire party was back a few days later. Zuwena's mother said that Ame had not been fulfilling his terms. Ame, however, said he had a letter from his *sheha* stating that Zuwena was not fulfilling her terms. Shaykh Hamid listened to them, then turned to Bwana Fumu, saying that he believed the parents were much to blame in the case. He said it was evident that Zuwena's father had detained her, but that technically Ame could not accuse him, because he was not "central" to the case, which was a dispute between a husband and a wife. Ideally, however, Ame should accuse the people who prevented his wife from returning to him. Bwana Fumu agreed that the matter was complicated and that the father seemed somewhat blameworthy.

Eventually, Shaykh Hamid ordered Zuwena to buy her divorce. He reminded them that Zuwena wanted a divorce, but her husband did not, and said that he did not like either Ame's or Zuwena's behavior in the marriage. He explained the laws governing *khuluu* and told them that even though the law would not permit exceeding Zuwena's *mahari* of 10,000 shillings, a man could not remarry for such a small sum, so negotiation was possible. Shaykh Hamid stressed that the problem was essentially between the fathers of the litigants. In addition, he thought that the fathers of both litigants had not tried to solve the problems of their children. Zuwena's father made a legal mistake by preventing her from going back to her husband, and the *kadhi* argued that this indeed was the reason they were permitted to negotiate the amount: Ame could demand more because of the mistakes made by Zuwena's parents. As we have seen elsewhere, Shaykh Hamid emphasized the role of elders in the lives of their children as both caretakers and moral guides. In this case, the parents' failings in this duty resulted in Zuwena buying her divorce.

Ame supported the negotiation of the *khuluu* amount, and said that he would give her a divorce for 150,000 shillings, which was 15 times her *mahari*. Zuwena's father was shocked, and said he would not pay this amount. Shaykh Hamid told them that if they could not agree on an amount together, then he would order Zuwena to return only her *mahari*. Eventually, the parties agreed on 50,000. Zuwena's mother brought the money to court a few days later, but when Ame came in to collect it he refused to take it. He complained that it was too little to remarry and he demanded another 100,000. Shaykh Hamid reprimanded Ame for his greed, and reminded him that they had agreed upon the sum. Ame left in a huff and did not return. When he was gone, Shaykh Hamid told me that Ame should have been grateful he allowed him to receive 50,000 since he had married Zuwena for only 10,000 shillings. In his written summary, Shaykh Hamid did not specify the amount of the *khuluu,* but recorded that the defendant refused to take the money because he wanted a sum that was far more than what he paid in *mahari.*

After listening to the plaintiff and the defendant and their witnesses and their desires, the court issues a decision. By following the laws set down in the Qur'an and the *hadith* of the Prophet and the verses concerning *khuluu* divorce in the *surah,* The Cow: *There will be no blame on either if the woman redeems herself* [14] and *Ibn Abbas' [narration concerning] the wife of Thabit ibn Qais.* [15] The defendant was given a divorce and the plaintiff refused it because he was not given 100,000 like he wanted even though he married her for only 10,000. Therefore, the case is closed, but the plaintiff has the right to appeal the case for one month.

Discussion

In the context of the Zanzibar courts, characterizing *khuluu* simply as a divorce by mutual consent or woman's initiative is simplistic. Although the women in these cases desired divorce, they did not come to court specifically asking for *khuluu,* and in two cases, were not the plaintiffs but rather the defendants. As we have seen, Shaykh Hamid used judicial *khuluu* as a rather powerful tool that functioned as both a compensatory and punitive measure. Moors describes a similarly punitive usage of *khul'* in Palestine: "If they were not able to reconcile the couple and the responsibility for the discord lay with the husband the court would divorce them, with the wife retaining all her rights. If, on the other hand, the wife was regarded as responsible the court would impose a *khul'* divorce, requiring the wife to renounce her dower and maintenance rights" (1995: 142). In her study of the emergence of *khul'* by judicial decree in twentieth-century Pakistan, Lucy Carroll notes that this "liberalization" of divorce law enables women to file for divorce on more grounds, but comes at a financial cost to women who forgo their financial claims upon their husbands in *khul'* she writes that "Judicial *khul'* is a dissolution with particular consequences, consequences that exonerate men and penalize women" (1996: 121). This might be said of the *khuluu* we have seen in Shaykh Hamid's court. Women often buy their divorces at very high prices and judicial *khuluu* is sometimes used as a "fine" for a woman's perceived bad behavior in a marriage. Although I could not follow through on all the cases, *khuluu* often causes financial hardship, and most women relied on their parents and extended families to help raise the money for it. However, we should also be careful to recognize that *khuluu* provides a means for women to divorce under almost any circumstances: although it may cost her significantly, a woman who does not want to stay married need not do so. Recall, for example, the way in which Shaykh Hamid wrote the conditions for *khuluu* at the end of every *masharti*: if a woman broke her prescribed terms, then she must "buy" her divorce in *khuluu*. This is similar to what Tucker has written about Ottoman courts, in that even though women might suffer financial difficulty as a result, "*khul'* represented a means by which a woman could initiate divorce proceedings if she found herself in an unwanted, although not legally defective, marriage" (1998: 97).

To use *khuluu* in a punitive fashion, Shaykh Hamid occasionally permitted actions that he thought were technically unlawful but appropriate in the local context. Several anthropologists have shown interest in exploring the relationship between custom, social norms, and understandings of Islamic law, and have questioned how culturally constructed notions of fairness and equity figure into judicial reasoning in diverse cultural contexts, Islamic and otherwise (Bowen 1998b, 2000, Just 1990, 2000, Rosen 2000, Jackson 2001, Peletz 2002). In his study of Malaysian Islamic courts, Peletz observed that judges were often more concerned with the public good than with narrowly applying legal rules (2002). Rosen ponders similar questions in Morocco, and addresses local notions of equity in considering how judges incorporate custom into their judicial reasoning when textual sources are wanting, or when the law is evident, but a *qadi* feels that applying the law would compromise justice (2000). In some situations, then, like Shaykh Hamid judges felt that application of the law would violate fairness.

In his work on legal reform and Islamic jurisprudence in present-day Indonesia, Bowen has argued that jurists have drawn on local ideas about fairness and equity to justify new codified rules about gift-giving and bequests (1998b, 2003). New laws in Indonesia have not gone the way of other Muslims states and other Indonesian reforms that attempt to bring Islamic law more in accord with local norms by relaxing rules about gifts and bequests. Rather, they made the processes more difficult (2003: 134). Interestingly, instead of simply applying the new rule, jurists have sought to explain how it corresponds with local ideas of fairness (1998, 2003). He argues that the Indonesian laws limiting bequests and gifts actually reflect historically developed norms of fairness, and that jurists later developed an Islamic justification for the rule that "weaves together general moral principles, implicit analogies, and reports of statements of the Prophet Muhammad. Social norms and religious reasoning generated a law-like norm, which subsequently received legal justification" (2003: 146).

In the Mkokotoni court, however, we see that in some cases Shaykh Hamid used fairness and the attribution of fault to explain why he occasionally allowed practices that were not entirely in accord with what he considered lawful. Shaykh Hamid took into consideration who was at fault, or at greater fault, in deciding on *khuluu* and in determining a fair sum. As we have seen, he sometimes allowed litigants to negotiate their own *khuluu* amount as long as they did so outside the walls of the courtroom. If the litigants agreed to the amount outside of court, it was not in egregious violation of his understanding of Islamic laws of *khuluu*. Shaykh Hamid explained that if the litigants reached an agreement themselves, he could accept it and even write it up in the court decision since he had not himself prescribed the amount in the court. This is somewhat similar to what Aharon Layish found in his work in Libya and Israel, where *qadi*s accommodated "voluntary agreements" made between litigants: "This attitude of the *qadi*s [in Libya] towards customary *khul'* is strikingly in line with the declared attitude of Muslim *qadi*s in Israel on the same issue, according to which the court does not interfere in voluntary agreement, holding that 'mutual agreement is stronger than the *qadi*'" (1988, 1975). However, Shaykh Hamid did not make the argument that mutual agreement was stronger than the *kadhi*, but rather that it was pragmatic in some cases. In 2002, when discussing yet another case in which a woman paid more in *khuluu* than she received in *mahari,* Shaykh Hamid described the situation as allowing a woman to achieve divorce when her husband refused:

> This man did not want to divorce her, so they increased [the amount] so he would divorce her. It's not the law, but when we gave the law [i.e., the court decision for *khuluu* at a lesser amount], the man absolutely did not want to divorce her, and she absolutely does not want to go back to him, and so therefore she said she would increase the amount, but it was not in the divorce decision.

They agreed on this amount, and they did not do it in here [the courtroom], they did it outside.

In other cases, he reasoned that women could pay more than they received in *mahari* out of fairness: if a wife had caused her husband undue trouble, she could be expected to pay a higher sum. Occasionally, he ruled that the higher amount was fair because it would allow a man to remarry and pay a new *mahari*. The burden of a financial responsibility did not only indicate who was at fault, but was thus the means of making right a situation and achieving equity. As we have seen in the cases of Mosa, Siri, and Ame, this most often took the shape of a financial obligation, but Shaykh Hamid used an alternative to cash in the case of Hamdu and Rukia. It should be noted, however, that even in situations in which he deemed the wife at fault, he would check a husband's greed if he thought he was asking for too much, as in the cases of Siri and Samir and Ame and Zuwena. We must recall, however, that Shaykh Hamid never condoned the variation of *khuluu* known as writing for money, specifically because it did not involve a woman's desire for divorce:

To write someone for money...it is not a right [that men have]. Absolutely not a right. The *sheria* says that to give money to the man who married you, you don't have the right to give him that money unless the woman does not want her husband. This is *khuluu*, and then the woman has the right. But if the man is there and demands money from his wife, that is not *haki* [not a right that he has]. But they do this. This is not *mila*. If I write that you will give me 1010 shillings and then I will give you a divorce, this is not *haki*. Unless the woman has refused her husband and said that she does not want him...then she gives me the money that she married me for. Now, that is the *haki* of the woman who paid the money because she paid after refusing her husband. But the husband does not have the right to demand money from his wife. Well, maybe if he borrowed it...but for marriage, this is not custom or religion. It is just a mistake.

There was variation in how *kadhi*s and other religious experts viewed *khuluu* and I do not propose Shaykh Hamid's approach was typical. As we saw in chapter five, Shaykh Hamid viewed his role as *kadhi* as one of compromise, and his views on *khuluu* seem to illustrate this. He knew that, as a state-appointed *kadhi*, he was not able to apply *sheria za* *dini* in full. Shaykh Vuai saw his position as state-appointed *kadhi* quite differently, and on more than one occasion he criticized Shaykh Hamid's practice of allowing *khuluu* negotiation as both inappropriate and unlawful. Even though Shaykh Vuai knew his jurisdiction was circumscribed by the state, he thought his greater obligation was to uphold proper interpretation and application of religious legal principles. Shaykh Vuai viewed *khuluu* as a legal process allowing women to divorce when their husbands did not desire it, and he told me on several occasions that he would not allow negotiation under any circumstances because paying more than *mahari* was strictly unlawful. He explained that by permitting negotiation of the *khuluu* amount, Shaykh Hamid was in effect condoning unlawful and harmful local practices like writing for money. Even if a husband and wife agreed on an amount outside of the courtroom, he insisted that a husband demand not a shilling more than what he paid in *mahari*. I recall that in one case, a young man asked this *kadhi* if he and his wife could divorce in *khuluu* when reconciliation failed. Shaykh Vuai responded by telling the young man that "money makes us drunk" and that a *khuluu* divorce was greedy and best avoided whenever possible, and he did not permit it.

A noted Qur'an teacher called Mwalimu Shabbani, who lived in a village a few miles from Mkokotoni, exemplifies a third position on *khuluu,* and provides an interesting contrast to the positions of Shaykhs Hamid and Vuai. When I asked him about *khuluu,* Mwalimu Shabbani said that if he were a *kadhi* he would allow negotiation of the amount and the payback of more than *mahari*. Unlike Shaykh Hamid, however, he gave a religious reason for doing so: he explained that the point of *khuluu* was to allow a man to remarry through returning enough money for him to pay a new *mahari*. Recall that Shaykh Hamid, in several cases, said that allowing a negotiation of *khuluu* would help the man remarry, but he did not explain this as the point of *khuluu,* but rather a consideration of fairness that was outside the lawful bounds of *khuluu*. According to Mwalimu Shabbani, however, it was religiously lawful to allow a man to receive more than he paid in *mahari* if the average amount of *mahari* had increased over the years. The issue thus centered on the proper interpretation of the meaning of the law, not on allowing a technically unlawful practice in some circumstances, as in Shaykh Hamid's reasoning.[16]

CHAPTER SEVEN

Conclusion: The Court Is a Hospital

Throughout the book, I have attempted to demonstrate how Zanzibari *kadhi*s, clerks, and litigants define, understand, and utilize Islamic law in one working court in one particular cultural context. Many states today make provisions for the application of Islamic family law. As in Zanzibar, it is often circumscribed by state law, and those who apply the law, like *kadhi*s, are beholden to the state legal system and, in many cases, a family law code. Although no code has been adopted in Zanzibar, it is possible that this will happen in the future. Also, as scholars have shown elsewhere in the world, Islamic judges may also refer to local norms or ideas that fall outside of or even contradict their understandings of Islamic law.

It is evident that Shaykh Hamid places great emphasis on preserving marriages through remedying marital strife, which seems to be a common aim of many judges in Islamic family courts around the world; historians and anthropologists have described other *kadhi*s in other courts who strive for the same thing (e.g., Agmon 2006, Peletz 2002). Like others in the surrounding community, Shaykh Hamid viewed the court as the last resort in a three-part process of dispute resolution. If the elders and *sheha* failed to solve the problems of a couple, then they brought the dispute to him. As we have seen, the *kadhi* often compared the court to a hospital, "Where do you go for *dawa* (medicine) for an illness? And if you are sick [with legal problems] you come to get your *dawa* at the court. You come and tell the *kadhi* that you are sick." A marital dispute was like a physical illness, and just like an illness could be treated and perhaps cured by doctors in the hospital, so a marital dispute was an illness that could be treated with the medicine the *kadhi* dispensed at the court. The emphasis on treatment and

cure was certainly reflected in the way in which he stressed reconciliation. Shaykh Hamid rarely ruled for an immediate dissolution of marriage, even in cases where reconciliation seemed somewhat hopeless.
The *kadhi*'s "medicine" most often took the shape of *masharti* that each
party must follow in order to achieve marital harmony. The *masharti*
and Shaykh Hamid's verbal instructions to the couple not only emphasized that they must uphold the duties required of husband and wife by
religious law, but also that they must strive to live together peacefully.
Peletz notes a similar emphasis in Malaysian Islamic courts, where *kadis*
try to "keep alive" failing marriages regardless of the specific aims of
husbands and wives (2002: 120). There is also a clear historical parallel
in the Ottoman courts of Palestine, where *qadis* aimed for compromise to serve the interests of both parties rather than simply make a
declarative decision about who was right and who was wrong (Agmon
2006).

Shaykh Hamid's emphasis on "curing" a problematic marriage through
reconciliation was evident in the frustration he expressed in one final case.
A young woman named Patima opened a claim against her husband,
Abduli, because he had not been supporting her for over eight months.
On their first day in court together, I arrived late and asked the *kadhi*
what was going on. He explained that he was annoyed because Abduli
wanted to divorce his wife through repudiation right then and there in
court. Shaykh Hamid told him firmly he must not divorce his wife because the court was "a place to solve problems—*not* a place to divorce!"
He continued by explaining that Patima had come to the court to demand her rights, and that they needed to proceed toward that objective
by continuing with the case. Despite this stern counsel, Abduli decided
to divorce Patima anyway, and there was nothing Shaykh Hamid could
do to stop him. This was one of the few times I saw the *kadhi* so angry.
He called Abduli's desire to divorce his wife a mistake and lectured
him at length about the duties a husband and wife have toward each
other. Although Patima did not seem terribly upset about the end of her
marriage, the *kadhi* told her to not sulk but to persevere and try to live
in peace with Abduli. He remained cross even after they left, and told
me again that Abduli had wanted to divorce her from the first moment
he arrived in court, which was *not* the way of the court. He had told
them to go home and think it over, but it still ended in divorce. "What
could I do?" he said regretfully, "*Sina njia ya kupita*" (I had no way to
do otherwise).

As we have seen, unlike in many other Muslim countries, men in
Zanzibar maintain the right to divorce by unilateral repudiation. Many

*kadhi*s and scholars expressed distaste for the seemingly cavalier way in which men sometimes used this right. Shaykh Hamid's comments about Abduli echoed his general feeling that repudiation must never be done without just cause. However, there was nothing he could do to stop it and ultimately, he believed, the matter was between the man and God. Of course, the *kadhi* issued no ruling in Patima's case and it would have been recorded in the register simply as "The defendant divorced his wife in court." However, he did make a final note in the case file, and though the comment was brief, his emphasis on Abduli's failure "to continue with the case" illustrates the potentially curative powers he believes the court holds for disputing couples, and his frustration at Abduli's rejection of his marriage.

> In the court of the *kadhi* at district of Mkokotoni, Unguja, the plaintiff and defendant are here. The defendant did not want to continue with the case and he wanted to divorce his wife, and he divorced his wife. One divorce. And the wife was written her *passi*. The case is finished today, February 29, 2000.

We might consider the emphasis on reconciliation in marriage to mirror Shaykh Hamid's general approach to navigating the waters between community, state, and religion in his work. As we have seen, the *kadhi* viewed his position as one of compromise. Dupret has noted that in Egyptian personal status cases, judges often pay little attention to what is specifically "Islamic" about that law. He writes of the "ordinariness" of a judge's routines, and describes a judge's work as legally characterizing the facts that are presented to him. He notes, however, "This does not mean that this or that legal provision has no history. Rather, it means that so long as a legal concept is used in a stable, unproblematic and unquestioned manner, any account of its theoretical and historical basis has no special relevance for its current uses" (2007: 86). This emphasis on "ordinariness" is similar to how Shaykh Hamid explained his own work. He rarely cited legal sources other than the Qur'an or *hadith* in his rulings, and he told me that this was because most of the marital disputes were so simple and formulaic that he did not need to consult or cite legal texts for explication. It was in those cases with more complicated legal issues at stake, as in determining intention in the alleged written repudiation of Mariam by Abdulmalik, that he cited legal texts.

As we have seen, Shaykh Hamid sometimes explained his decisions as in accord with *sheria za dini* and sometimes he used other means to justify his rulings, and even sometimes allowed arrangements in rulings that he considered specifically unlawful. He justified such rulings in different ways. Sometimes, he allowed "unlawful" resolutions if they occurred outside of the court at the agreement of the parties involved, and in these situations, the court did not technically give the unlawful ruling. At other times, however, what was technically unlawful was a basis for a particular ruling. Drawing on local notions of what constitutes appropriate marital behavior, he justified such decisions as being in the interest of fairness or making right a difficult situation. Sometimes this was achieved through punishing the guilty party by making them pay, or by arguing that coming up with another *mahari* presented a difficulty for a man hoping to remarry after a divorce. As noted in chapter six, he neither used overarching Islamic legal principles to justify such rulings nor appealed to permissibility of custom in Islamic law. Rather, he simply explained that they were unlawful but appropriate under the circumstances.

Rulings like this reflect the way this *kadhi* understood his position as one of compromise. On several occasions, he told me that others rejected the position of *kadhi* because they could not apply religious law in full, "They refused because of the use of the law. You see, there are laws that must be fulfilled, and the *kadhi* does not have permission to fulfill these laws." I asked for further explanation, and he complied by giving several examples,

> Like the laws concerning drunkenness [for example]. If a person gets drunk, there is a law you must use according to religion. But you cannot follow it because, if you do, you will be going behind the back of the *kanuni* (state law). And if a *kadhi* wants to do that [neglect *kanuni*], then he can't do the job. Another example is theft. If a person is a thief, you have to cut the hand off [according to religious law], and people will say that if they cannot rule like this [i.e., apply the law in full] then they do not want the job. [And with] the law of adultery, you are supposed to beat the offender, but the government says you can't do that.

We continued along these lines another day when talking about a case involving a young woman called Hamida, who had come to court asking for permission to marry her boyfriend. Her father and his kin refused to permit the marriage, even though Hamida was already pregnant by the young man. As noted in chapter one, pregnancy out of wedlock is technically a criminal offense, though cases are rarely brought forward and most people of my acquaintance were pragmatic about the matter: such pregnancies are considered unfortunate, but a simple fact of life, and young women in these circumstances were usually supported by their families. For example, in my last visit in 2008, Mzee Bweni's granddaughter, about 23, had recently had her second child out of wedlock, and though the family was certainly disappointed, they loved and supported her as they always had. Hamida's situation thus technically involved an offense, but Shaykh Hamid wanted to avoid going to the secular court even though, as he told me, the penalities for pregnancy out of wedlock in Zanzibar were much lighter than in classical Islamic law. "This [Zanzibar law] is not the law in the book," he said as he tapped the well-worn Qur'an on his desk with a pen. He explained that in "the book," the law was like that followed in Iran or Saudi Arabia, which involved corporal punishment. He paused, and then with a thoughtful expression in his eyes, explained that such laws were "a bit too heavy" for Africa and thus not appropriate. In comparison, he said, the Zanzibar law was "very light." In sum, according to Shaykh Hamid, application of "full" Islamic law was not possible in Zanzibar and was also not appropriate, as with this example. Despite the "lightness" of the law, he still wanted to avoid involving the criminal court in Hamida's case. To the *kadhi*, the ideal resolution would be the marriage of the two young lovers, not punishment for their affair, and he eventually wrote a ruling in which he argued that a father could not prevent an adult woman from marrying. He married the couple himself.

Court-based *kadhi*s are government appointments, but their religious function makes them somewhat different from other governmental employees and secular judges. The relationship between courts, community, and states has been an important area of scholarship, particularly in considering the role of the court as arena of local politics between community and state (e.g., Starr 1990, Lazarus-Black and Hirsch 1994, Reiter 1997, Wurth 1997, Peletz 2002), and in tracking changes in legal reasoning (e.g., Layish 1975, Messick 1993, Shaham 1995, Bowen 2003). Many scholars have looked at courts and dispute resolution processes as arenas in which to affirm or contest community values and politics (e.g., Lazarus-Black and Hirsch 1994). In their present-day form, the Zanzibari courts and their *kadhi*s are creations of the state, which drew on legal reforms introduced in the colonial period that modified Islamic procedural law and limited jurisdiction. Today, *kadhi*s are appointed and paid by the state. However, I propose that it is simplistic to view today's *kadhi*'s courts as simply an arm of the state.

Shaykh Hamid was clearly interested in the implications of his position between the community and the state, and he perceived a responsibility to both, as well as to Islam. This is evident in the way he incorporated different voices of authority into court proceedings through witnessing. Although he claimed to be unaware of state requirements of witnessing, he was well aware of the role of the *sheha*s in dispute resolution, and emphasized this part of proper procedure in most cases by requiring litigants to see *sheha*s and by inviting them to testify. He also accommodated local modes of scholarly authority in the way in which he incorporated the expertise of local *kadhi*s and the participation of elders into the proceedings.

Certainly, the state has had significant impact on the shape of the courts and their jurisdiction, and this will likely increase in the future. As described in chapter one, several reforms have been implemented since my initial research trip in 1999 and 2000. Zanzibar has suffered a tense political climate for many years, with highly contested elections and many allegations of fraud in both the 2000 and 2005 elections, in which the ruling party, CCM, was declared victorious. Political awareness and anxiety is high even among men and women in the rural areas, and the opposition is sometimes associated with religious leadership and Islamic activism. When discussing my research, people sometimes wondered aloud whether people working in the government were appropriately religious, and occasionally, though not frequently, these comments implicated the religiosity and learning of the state *kadhi*s. Although Purpura found negative attitudes about *kadhi*s in Zanzibar Town (1997) and Hirsch noted routine criticism of Kenyan *kadhi*s for their association with the state (1998), I rarely heard negative political associations with Shaykh Hamid or other *kadhi*s, although I certainly heard this about other state appointments. Furthermore, although no one told me that they would avoid the *kadhi* because he was a political appointment, people occasionally wondered if the *kadhi*s' link to the state compromised their ability to apply religious law in full, which, of course, was the opinion of Shaykh Hamid himself.

The subject of women and Muslim family law has received a great deal of attention from scholars in recent years. In contrast to earlier work on gender on the Swahili coast, I have found no indication that people in rural Unguja view *sheria* as the domain of men more so than women. Like Bi Mboja said in chapter four, "there is no stranger to marriage" in Zanzibar. We have seen that cases involving disputes over whether a divorce occurred do not stem from women's lack of religious and legal knowledge, but from the events they experience during divorce. And, as we have seen, the way in which women present claims in court often show great familiarity and facility with Islamic law concerning marriage and divorce; if anything, they were perhaps more savvy than men. Consider again Fatuma's insistence that her mother deemphasize that she wanted a divorce in an attempt to avoid *khuluu,* how Zaynab and her father focused together on Rashidi's abuse as grounds for "free" *fasikhi* divorce, and how Mariam taught Abdulmalik to divorce her. In this book, I could touch only lightly on how women and men acquire religious and legal knowledge in Zanzibar, and this appears to be a promising area for future research.

As noted in chapter one, a number of scholars working on gender and Islamic courts have proposed that, contrary to popular conceptions, Muslim women often fare well in courts. *Kadhi*s in Zanzibar regularly emphasized the importance of helping women in their work. Shaykh Vuai once told me that he enjoyed his work as a *kadhi* because he was fond of women, and wanted to help them. He also said that this was proved problematic sometimes, because he was occasionally overly sympathetic to women. Shaykh Faki lamented that women, especially in rural areas, suffered greatly at the hands of men who often violated the religious laws protecting the rights of women. Does this mean that women win more cases than men do in Zanzibar courts? Is the court truly "one-eyed" as Mzee Bweni claimed? Not exactly, because men often won claims in court. However, women were less likely than men to lose cases, and in those that were "settled," women often achieved some of their aims, and rarely left empty-handed. As we have seen, however, it is somewhat difficult to even establish who won a particular a case. Once we consider these cases in their entirety, there is rarely a clear "winner" or "loser" regardless of who actually opened the case in the court. Describing cases with reference to who won and who lost simplifies what is happening in the court—this is true even of cases that ended with a ruling rather than settlement. Although useful to some degree, counting cases as wins or losses has its limits. In many cases, a final decision often came about through a great deal of negotiation and many manifestations, and some cases had two decisions recorded. Take, for example, the case of Zaynab and Rashidi. Zaynab came to court claiming that her abusive husband had repudiated her. The alleged out of court divorce was never established, and as in so many other cases, the *kadhi* ruled for reconciliation. According to the *masharti,* Zaynab's first requirement was to return to her husband. She did not do this, so bought her divorce in *khuluu.*

Therefore, she incurred a significant financial hardship, but got her divorce. Did Zaynab win or lose?

As we have seen, Shaykh Hamid rarely ruled for an immediate divorce, instead hoping to preserve a marriage through the medicine of *masharti.* However, litigants could always opt out of taking this "medicine." Every ruling of this sort stated that if either party broke the terms of the contract, it would result in a divorce of a type of most benefit of the aggrieved party. As a result, a woman could always get a divorce provided that she is willing to pay for it. She merely had to break the *masharti* and no one would force her to return to her husband's home; we saw this in Rukia's case.[1] Furthermore, women nearly always have the support of some if not all of her elders and family members. In all my interviews in court, I asked women about family support: none indicated they were there against family wishes and most came to court with family members. So perhaps Mzee Bweni was right in one regard: a woman who wants a divorce can always do so via the court, even if she has to buy it. Despite this, it should be noted that most divorces happen outside of court, and most women (and men) never take a dispute to the court. Even though some women appreciate the relative ease of taking disputes to court these days, others choose not to take grievances to court because they are reticent to air problems in public, as we saw with Mwanahawa in chapter three.

Between 1999 and 2008, I conducted about two years of research in Zanzibar, and most of that time was spent with Shaykh Hamid, Bwana Fumu, and the others in the Mkokotoni court. When I returned to Zanzibar in 2002 for my second field trip and my last while Shaykh Hamid was alive, I found him working with a new *kadhi* who eventually took his place at the court. They were sitting side by side at his desk, and reviewing a case file. After our initial cheery greetings, Shaykh Hamid told the more reserved new *kadhi,* Shaykh Zubeiri, that I had come to Zanzibar years ago to study "the laws of religion." He said that I was sent by my elders and my university and that after staying for a very long time, I went home and successfully completed my studies. He told Shaykh Zubeiri that we had talked about many, many things during my stay—local dialects of Kiswahili, the Qur'an, marriage, and divorce. He described our methods just how I would have done so—I listened, asked questions, and got translations of things I did not understand.

Shaykh Hamid's enthusiasm was essential to my work. I am sure it easily opened many doors that would have otherwise been closed. Shaykh Hamid and Bwana Fumu were supportive of my research throughout and, just as he did with Shaykh Zubeiri, the *kadhi* always took great care when explaining to visitors my presence in the court. He would describe me as a student of Kiswahili and *sheria,* and stressed the importance of me taking my newfound knowledge back to America to educate others about this court and life in Zanzibar. I hope I have accomplished this to some degree.

NOTES

One   *Kadhi*, Court, and Anthropologist

1. For a comparative look at the status of Islamic family law in different parts of the Muslim world today, refer to Abdullahi an-Na'im's overview of Islamic family law (2003).
2. See Susan Hirsch's article (2006) for a thorough overview of sociolegal scholarship on Islamic law.
3. There are a number of different schools of Islamic legal thought that are named after the influential thinkers behind the development of the schools. Of the four Sunni schools, Shafi'i is the most prominent in East Africa. The Omani Arab colonizers were Ibadhis, and a few Zanzibaris still identify as Ibadhi. There is also a small Ismaili Bohora community in Zanzibari Town.
4. Women in Zanzibar are addressed with the title *Bibi* in front of their name (*Bi*, the short form, is more common), which means Ms. The title is used for women of all ages and marital status. For younger men, the title *Bwana* (Mister) is used and for older men, the title *Mzee* (Elder) is used.
5. For a thorough treatment of the recent history of the *kadhi's* courts until 1963, see Stockreiter's dissertation (2008).
6. Many scholars have noted a preference for oral than written evidence in Islamic legal context, (Messick 1993). Stockreiter traces the influence of the colonial period in establishing a reliance on written evidence (2008), and in Zanzibar courts today, written evidence is much preferred in many situations.
7. This period also saw other strict laws in Zanzibar concerning dress and hairstyle, like the 1973 decree on clothing, and placed strict regulations on visits to Zanzibar from foreigners. See Burgess (2002).
8. This is the Act to Repeal and Replace the Protection for Spinsters Decree and to Provide for Protection of Widows and Female Divorcees and to Provide for Other Matters Connected Therewith and Incidental Thereto, No. 4.
9. This is the Spinsters and Single Parent Child Protection Act, No. 4 (2005).
10. The bill is entitled "The Amendment Concerning a Law to Establish the Office of the *Mufti* of Zanzibar and Matters Concerning the Establishment Thereof" (No. 93 of 2001).
11. Marital disputes in which abuse is the primary complaint are handled as secular criminal cases; those inheritance disputes in which the problem concerns property itself rather than determining proper heirs or the proper division of property are handled in the secular courts.
12. Divorce is also common elsewhere on the Swahili coast (Strobel 1979, Landberg 1986, Mirza and Strobel 1989, Middleton 1992, Hirsch 1994, Caplan 2000).

13. I modeled this table after a similar one in Susan Hirsch's *Pronouncing and Persevering* (1998).

Two   Writing a Case: Court Actors and Court Procedure

1. As of 2008, fashions were changing, and some women in town (and when going to town) were alternating the *buibui* with an elegant outfit known as *kitanga* (in the fashion of Tanga, a coastal city on the mainland). This new style consisted of a colorful long dress and matching scarf that covered the head and upper body in a way similar to a *kanga*, but more was elegant and made with more expensive fabric. It was thus quite different from the black *buibui*, but covered the body in a similarly modest way.
2. Although several people at home in the United States often asked me how I was received as a woman working in such a court, no one in Zanzibar's courts ever commented on my gender or seemed to think it of any importance at all.
3. Shaykh Hamid listed the shaykhs after whom three of the Sunni legal schools are named.
4. In a 2005, one *kadhi* told me he had originally been offered the position several years back, and turned it down because he was "tired" and did not want the hassle of the job. He eventually took the position.
5. Prior to the twentieth century, Kiswahili was written in Arabic script, and even people born in the twentieth century learned to read and write Kiswahili in Arabic script. As a result, some *kadhi*s, Shaykh Hamid among them, had to learn to write Kiswahili in the Roman alphabet when they were appointed to their positions.
6. In these situations when they share an office and courtroom, the *kadhi*s and judges come to court on alternate days.
7. In their work on American courts, Yngvesson and Merry have noted the important role of clerks in shaping cases. Yngvesson proposes that the clerks work as "gatekeepers" in their power to manipulate the law, define problems as legal, and in some cases even "create" crimes (1993). In Merry's analysis of working-class legal consciousness and culture in the United States, she similarly describes the way in which the various parties to the disputes frame, categorize, and typify cases according to the legal and nonlegal categories they have developed (1990).
8. Agmon notes that keeping records of the court proceedings was the main duty of the scribe in the Haifa and Jaffa courts; their role was thus somewhat different from the clerks in Zanzibar's courts.
9. The role of the *sheha* will also be addressed in chapter five.
10. Because plaintiffs were most often female, I use the female pronoun here. However, the process was the same regardless of the gender of plaintiff and defendant.
11. Stockreiter notes that Nawawi's *Minhaj al-talibin* and the corresponding commentary by Ibn Hajjar al-Haytami were main references for Shafi'i *kadhi*s in the early to mid-twentieth century (2008: 75).
12. To protect their privacy, the names of all litigants are changed throughout the book; details about home villages and *shehia*s are also concealed.
13. Machano's ethnicity is listed as "Mtumbatu." Most people in this area claimed to be Tumbatu, but sometimes people simply answered "Zanzibari" or "Swahili." Occasionally, litigants were confused when asked this question, and I remember one young woman telling Bwana Fumu to just write down whatever he was.
14. In the Zanzibari calendar, the month of *mrisho* is Ramadan, and *mfunguo mosi* is Shawwal.
15. Qur'an 2:229. The *kadhi* always cited the Qur'an in Arabic; I use Ahmed Ali's translation throughout the book.

16. Shaykh Hamid often tried to determine the nature of a divorce even when the litigants or clerks prioritized another legal issue. This is due to the frequency of the practice known as "writing for money," which Shaykh Hamid and other legal scholars in Zanzibar consider a misinterpretation of Islamic divorce.

17. The "inability" to get along with their husbands was a frequent complaint of women coming to court.

18. Much debate ensued about how much she would pay for her divorce, but this will be taken up in chapter six.

Three   From Community to Court: Gendered Experience of Divorce

1. See table 1. 3. Although marriages and divorces are supposed to be registered with government authorities, few divorces are reported in a timely manner, if at all. Legally, an official receipt of divorce called must be presented to an official before a new marriage can be authorized. Despite the fact that some officials require only a woman's word that she is divorced, the power of the divorce receipt is widely recognized. In a sample of cases from the first half of the twentieth century in Zanzibar Town, Stockreiter notes that 7 of 206 cases involved similar request for confirmation of divorce (2008: 102).

2. Radio Zanzibar was popular during much of my field work, and the station scheduled several programs of a religious nature that address family life, including women's rights and obligations in marriage according to Islamic law. Many women who owned radios would listen all day. Babu discusses the significance of the radio as instrument of education after the 1964 revolution (1991).

3. I have not done a close study of the economics of marriage, but note that this is a promising area for future research.

4. Following the revolution, wide-scale land reforms by the revolutionary government stripped Arab and Indian landowners of their properties with the goal of redistributing to "native" Zanzibaris; see Christina Jones-Pauly (1997). Since the issue never came to the fore in marital dispute cases, it suffices to say here that at present, the government technically owns most land, and Zanzibari citizens are given use rights to plots upon demand. They do not pay any rent for the land, and have use rights over trees, plants, and produce. Although I initially expected to witness arguments over land and usufructory rights to surface in marital disputes, this did not happen; joint property in marriage is generally not recognized.

5. For detailed descriptions of marriage elsewhere on the Swahili coast see Hirsch 1998, Caplan 1995, Middleton 1993, Mirza and Strobel 1989.

6. Adolescent girls are educated about womanhood by a *somo*, an older female relative who advises her until her marriage. Mwanahawa often joked that I would be her *mwari* when I got married, and she would be my *somo*.

7. Hirsch has noted parallels in Mombasa, and the utility of the *buibui* as a disguise for girls wanting to attend weddings and behave in rowdy fashion (1998: 51).

8. Middleton (1992) describes a similar emphasis on education elsewhere along the Swahili coast.

9. Polygyny is fairly common here, even among men with modest means (see also Nuotio 2006).

10. This is also common on Mafia Island (Caplan 1984).

11. Some *kadhi*s, scholars and other professionals consider the norm of accepting a suitor without question a problem. Asia, who worked for UNICEF, thought that women would be better

off if they lived more in accordance with Islamic law and if women knew their rights under it. She was concerned that by accepting proposals without question, women were forced to comply with undesirable marriages.

12. Women often express a modest reluctance to marry, even when eager to do so.

13. The statement is often simple, such as "I, Juma, divorce you, Tatu." Some add "You are no longer my wife" or the number of divorces "I divorce you one time." In conversation, the verb *kuacha* (to leave) is also used for divorce. For example, a woman may say *nimeshachwa* (I have already been divorced).

14. Tucker noted such problems in Ottoman courts, where judges had to determine if a repudiation was binding or whether it occurred at all (1998).

15. Other anthropologists have noted Islamic judges' sympathy, or perceived sympathy, for women in courts (Hirsch 1998, Mir-Hosseini 2000).

16. Bwana Fumu was referring to deferred dower, when a marriage contract states that a man will defer paying some *mahari* until (if) he divorces her. I saw very few instances of deferred dower in northern Unguja, though Stockreiter notes it in early twentieth-century Zanzibar Town (2008).

17. Middleton described men asking their wives to return bride wealth in various Swahili communities when they initiated the divorce: "most divorces take place in the dry season, when there is less money about and men can claim back their bride wealth if they initiate divorce" (1992: 127). Aharon Layish notes similar variations on *khul'* in North Africa (1988).

18. This is an example of the ambiguity of the term *mila*.

19. Women usually return to their mothers to give birth.

20. As we see with Shindano and Zaynab, women may note the lack of maintenance in conjunction with the structural events as a further indicator of divorce. However, women did not specify lack of maintenance as a primary indicator because men often fail to maintain their wives even when there is no suspicion of divorce.

21. In one skit, the "husband" chastised his "wife" for cavorting with friends. She replied, "Don't you want them to invite me to weddings?" This emphasis on wedding attendance—a major social activity for women—reflects Hirsch's critique of Swartz's view that Swahili women dress up and attend weddings simply to show off (1982); rather, she argues that weddings are an important way for married couples to cement status in the community (1998: 98).

22. Hirsch noted that friends are often considered to "bring trouble" to a marriage through talk (1998: 66).

23. Calling someone a "dog" is a common insult. There is a marked contrast to what Hirsch found in Kenya's *kadhi* courts, where abusive speech was limited by the *kadhi* because it was considered harmful to those hearing it (1998: 214–218, 232). In Mkokotoni, reports of abusive speech are considered important and recorded in documents. Although a notable difference, the contexts show a similar concern for the power of abusive speech, and illustrate different means of handling the dangers of abusive speech. All marital abuse cases, verbal or physical, are criminal matters and thus under the jurisdiction of the civil courts in Zanzibar. However, in situations in which verbal abuse is one complaint among many, the case was handled by the *kadhi*.

24. Although physically beautiful, life seemed more difficult in this region. In recent decades, however, women have engaged in farming seaweed in the shallow coastal waters for export.

25. Sura IV.

26. The reasons behind Zaynab's *khuluu* amount are discussed in chapter six.

27. This is highly unusual. Technically, the Chief *Kadhi* only hears appeals cases. I was never certain why Rajabu went to the Chief *Kadhi* before a primary *kadhi*, or why the Chief *Kadhi* heard his case. This was also the only case in which I heard of a woman being forced to return to a husband.

28. Caplan noted a similar ambiguity about "proper" divorce among women on Mafia (1997).

Four   A Wily Wife and a Headstrong Husband: Determining Intention

1. For discussions, see Schacht (1964), Rosen (1984), Wensinck (1987), Esposito (1982), Baktiar (1996), and Messick (1998, 2001).
2. People rarely use a simple "no" (*hapana*) or a hard past negative in favor of a the *ja*- tense that, while negative, is more akin to saying "not yet" and thus leaves open the possibility for future action.
3. Bowen argues that the Gayo of Indonesia do not presume to be able to discern the inner states of other people. To demonstrate, he considers twentieth-century religious debates about announcing intention in the ritual acts of prayer. The debates focused on the meaning of ritual practice was the rite about communication with God, or about obeying God's commands? The central issue under dispute was whether ritual actions have built-in intentions. Gayo modernists argued that declaring intent in worship could weaken the idea that actions have built-in intent by implying that a declaration of intent is necessary (1999: 165). Traditionalist scholars countered by acknowledging that intent determines the value of an act, but that it would be impossible to know the true intent of another person—so he or she should not be criticized.
4. At this point, the *kadhi* announced that he wanted to discuss the case with the Chief *Kadhi*. However, when they all met again, Shaykh Hamid did not mention a meeting with the Chief *Kadhi*, and it was not clear whether he had gone, and I forgot to ask him.
5. In this case, Shaykh Hamid cites the Shafi'i legal text *Bughyat al-Mustarshidin*, a collection of *fatwa*s compiled by the Hadrami scholar Abd al-Rahman Muhammad Ba Alawi (1952); elsewhere (Stiles 2003), I mistakenly attributed this collection to another compiler. There is a long history of contact and collaboration between East Africa and Hadramawt region of Yemen, and Hadrami scholars have been influential in East Africa (see Anne Bang 2003). See Wael Hallaq for a discussion of the influence of muftis and fatwas on substantive Islamic law (1994).
6. Stockreiter writes about similar cases in Zanzibar Town in 1921; the *kadhi* Shaykh Tahir, however, determined that the writing of the divorce "was necessarily an expression of intention" (2008: 247).
7. I was not able to copy the court documents for this case, so I do not include translations. I was, however, able to read the documents and I summarize them when necessary.
8. A *talaka* is revocable after the first or second pronouncement; this means a man and woman can reconcile and return to their married state. After the third pronouncement, the divorce is irrevocable. *Khuluu* is irrevocable.
9. Qur'an IV, *Surat al-Baqarah*.
10. Cowives living together is extremely rare in Zanzibar—this was the only case I encountered where they did so.
11. Later, my housemate Mwanahawa, who was from the same village, told me that of course the *kadhi* acted with great deference: he knew their status and would have made a grave mistake if he had not sought their advice on the case.
12. I do not propose that this is typical of all Zanzibari *kadhi*s. Shaykh Vuai told me that his university education in Saudi Arabia and his focus on psychology there gave him the ability to "see directly into people's hearts and to know their motives."

Five   Witness and Authority: Elders, *Sheha*s, and Shaykhs

1. Qur'an 4:35.
2. This amended the Local Government Act of 1986.

3. Although they are most often men, there are a few female *sheha*s on Pemba and Unguja.
4. Magnus Echtler has noted that in southern Unguja, one position of *sheha* had been passed through three generations of a family, though he does not note if this is typical in the region (2006: 150).
5. The concept of "insider" knowledge came up in discussions with others about language, marriage, and habits; for example, people use the phrase *Kiswahili cha ndani,* the Swahili of the inside, to refer to colloquialisms that might not be understood by outsiders. Hirsch notes a similar emphasis on "insider" knowledge in her study of courts in coastal Kenya (1998: 13).
6. The section reads as follows: The laws and rules of evidence to be applied in the *Kadhis* Courts including that of a Chief *Kadhi* shall be those applicable under Muslim law. Provided that—i(1.) all witnesses called shall be heard without discrimination on grounds of religion, sex, or otherwise; ii(2). each issue of fact decided upon an assessment of the credibility of all evidence before the court and not upon the number of witnesses who have given evidence; iii(3). no finding, decree or order of the court shall be; iv. reversed or altered on appeal or revision on account of the application of the law or rules of evidence applicable or rules of evidence applicable in the High Court, unless such application has in fact occasioned a failure of justice.
7. Anna Wurth found the same among women in Sana'a, Yemen, though men were also likely to bring work colleagues or friends (1995).
8. Despite this understanding, the *kadhi* occasionally required women to perform conventional gendered duties in written *masharti*.
9. Failure to live up to the terms given by the *kadhi* constitutes a criminal offense—for example, not paying back maintenance when ordered. However, I never saw anyone tried for this.
10. Unfortunately, I was not in court on the day he read the decision, and I do not know how they reacted.
11. Shaykh Hamid did not cite a source for this excerpt.
12. It was unusual for a woman to wear mismatched *kanga*s when out in public. Even the poorest women usually have at least one matching set. It is possible that Mpaji wore this to emphasize her poverty at the hands of a husband who would not buy her new clothing, but I was not certain.
13. Qur'an 4:34. Scholars have translated this verse into English in many ways. For consistency with the rest of the book, I have chosen Ahmed Ali's translation (1993).
14. This is in stark contrast to what Seng has described in sixteenth-century courts in Istanbul where records show that it was floating Ottoman *kadhi*s who were more likely to rely on local authority since they did not have the same knowledge of community matters (1994: 185).
15. There is a parallel in what Stockreiter has found about earlier *kadhi*s in Zanzibar Town. Writing against historians who assumed the *kadhi* was a collaborator with the colonial government, she notes that "judgments from 1900 to 1963 show that they essentially preserved their autonomy in the realm of family law" even though they were beholden to both colonial government and community (2008: 43).

Six Buying Divorce through *Khuluu*

1. Ahmed Ali translation.
2. Although money changed hands, these were not disputes over property. At most, a woman might complain in court that her husband tried to take away her clothing or *vyombo* out of spite. The only property I ever saw claimed in a divorce case was a bicycle, and both parties

agreed that the wife, the plaintiff, owned it, and had lent it to her husband, the defendant. She requested its return upon their divorce and her husband eventually returned it via the court; the seat was missing, and he was ordered to pay for it.

3. *Istihsan*, a legal principle concerning fairness that is similar to the Western legal concept of equity, might have been relevant; Mohammad Hashim Kamali writes that the difference between the two is that *istihsan* is based on the "underlying values and principles of *shari'ah*," while equity relies on the idea of natural law (1991: 245). Kamali describes the academic controversy between Hanafi and Shafi'i jurists over the place of *istihsan* in jurisprudence, but concludes by acknowledging the gap between legal theory and practice and that *istihsan* has been little used in the day-to-day adjudication. He finds the latter unfortunate, "especially in view of the eminent suitability of *istihsan* in the search for fair and equitable solutions" (264).

4. Despite this, we should not overlook regional and historical variations in Muslim *mahr* practice and its social and economic implications. Moors, for one, has critiqued Jack Goody's earlier work characterizing *mahr* as "indirect dowry" (1973) for overlooking regional and historical differences in *mahr* practices and for assuming that the *mahr* property forms a joint fund for the newly married couple (1996). More recently, she examined changing *mahr* practice in the socioeconomic contexts of urban and rural women in Nablus, Palestine, and argued that the practice of registering a "token" dower is the result of "different trajectories toward modernity" (2008).

5. Similarly, Hirsch found that Swahili women made few claims for *mahari* in coastal Kenyan *kadhis'* courts (1998).

6. For comparison purposes, the average teacher's monthly salary at the time was about 30,000 shillings (45 USD at the time).

7. Hirsch notes this among Swahili families in coastal Kenya (1998: 300), and Moors describes something similar in Palestine, although it seemed to happen more often in the past: "Although at present brides generally receive the full dower themselves, this was not always the case in the rural areas. In the past a village bride's father only gave part of the dower to his daughter and the father's share functioned as a (limited) circulating fund" (1995: 149).

8. Caplan describes a similar practice in Mafia (1984).

9. Among wealthier families in town, "kitchen parties" are popular. Somewhat like an American bridal shower but often fancier, women friends of the bride and her mother will bring gifts for use in the married home.

10. I wonder if this might explain some cases of "writing for money." Some men may understand *khuluu* as possible when they deem a wife has caused problems in the marriage, and thus could be using this idea of fault to reason that a blameworthy woman should buy a divorce even if the husband initiates it.

11. Like many other men, Samir referred to the *khuluu* money as "his" money. The *khuluu* is usually understood to be a return of the *mahari* money, and it is notable to many men seem to regard this as their money even years after it was given to a bride.

12. Rukia's age is listed at 36 years old, and the length of their marriage at 26 years. This does not mean she married at age 10. Many Zanzibaris do not keep track of their age in years, and court documents simply reflect the age they report to the court clerks. Rukia may have about 50, and her husband somewhat older.

13. At this point, the *kadhi* announced he would call witnesses, though I was not in court the day they came.

14. Qur'an 2:229; Ahmed Ali translation.

15. Shaykh Hamid wrote this line in Arabic, too, which references the *hadith* narrated by Ibn Abbas about the wife of Thabit ibn Qais asking the prophet about divorce. I hand-copied this portion of the text from the written *hukumu*, and afterwards could not make out the word following "Ibn Abbas." This *hadith* is used in support of *khuluu*.

16. His take on the matter seems to resemble most closely the way in which Rosen reports on the relationship between custom and religion in Morocco (Rosen 1995).

Seven   Conclusion: The Court Is a Hospital

1. Indeed, in only one case, "Jabu's Two Husbands" presented in chapter three, did I hear of a woman being forcibly taken to her husband's home. In Jabu's case, it was after that anomalous ruling by the Chief *Kadhi* that she was taken back—not as a result of Shaykh Hamid's decision.

BIBLIOGRAPHY

Agmon, Iris (1996) "Muslim Women in the Court According to the *Sijill* of Late Ottoman Jaffa and Haifa: Some Methodological Notes." In Amira el Azhary Sonbol, ed., *Women, the Family, and Divorce Laws in Islamic History*. Syracuse: Syracuse University Press: 126–142.

——— (2006) *Family and Court: Legal Culture and Modernity in Late Ottoman Palestine*. Syracuse: Syracuse University Press.

Ahmed, Leila (1992) *Women and Gender in Islam: Historical Roots of a Modern Debate*. New Haven: Yale University Press.

Allen, James de Vere (1993) *Swahili Origins*. London: James Currey and Ohio University Press.

Allen, James de Vere and Thomas H. Wilson, eds. (1982) *From Zinj to Zanzibar: Studies in History, Trade and Society on the Eastern Coast of Africa*. Frankfurt: Frobenius-Gesellschaft.

Allott, Anthony (1976) *The Development of the East African Legal Systems During the Colonial Period*. In Low, D.A. and Alison Smith, eds., *History of East Africa*, vol. 3. Oxford: Clarendon Press: 348–382.

Alpers, Edward (1984) "Ordinary Household Chores: Ritual and Power in a 19th-Century Swahili Women's Spirit Possession Cult." *International Journal of African Historical Studies* 17(4): 677–702.

Anderson, J.N.D. (1955, 1970) *Islamic Law in Africa*. Oxford: Oxford University Press.

An-Na'im, Abdullahi (2003) *Islamic Family Law in a Changing World: A Global Resource Book*. London: Zed Books.

Antoun, Richard (1990) "Litigant Strategies in an Islamic Court in Jordan." In Daisy Dwyer, ed., *Law and Islam in the Middle East*. New York: Bergin and Garvey: 35–60.

Asad, Talal (1986) *The Idea of an Anthropology of Islam*. Washington, DC: Georgetown University.

——— (1993) *Genealogies of Religion: Discipline and Reasons of Power in Christianity and Islam*. Baltimore: Johns Hopkins University Press.

Askew, Kelly M. (1999) "Female Circles and Male Lines: Gender Dynamics on the Swahili Coast." *Africa Today* 46(3): 67–101.

Association of NGOs in Zanzibar (2007) "MKUZA, A Plain Language Guide: Zanzibar Strategy for Growth and the Reduction of Poverty."

Austin, J.L. (1975) *How to Do Things with Words*. Cambridge, MA: Harvard University Press.

Ayany, Samuel G. (1970) *A History of Zanzibar: A Study in Constitutional Development 1934–1964*. Nairobi: East African Literature Bureau.

Ba Alawi, Abdul Rahman b. Muhammad (1952) *Bughyat al-Mustarshidin*. Cairo: Mustafa Babi al-Halabi.

Babu, Abdulrahman Mohamed (1991) "The 1964 Revolution: Lumpen or Vanguard?" In Abdul Sheriff and Ed Ferguson, eds., *Zanzibar Under Colonial Rule*. London: James Currey: Ohio University Press: 220–248.

Bang, Anne K. (2003) *Sufis and Scholars of the Sea: Family Networks in East Africa, 1860–1925.* London and New York: Routledge.

Bierwagen, Rainer Michael and Chris Maina Peter (1989) "Administration of Justice in Tanzania and Zanzibar: A Comparison of Two Judicial Systems in One Country." *International and Comparative Law Quarterly* 38: 395–412.

Bissell, William C. (1999) "City of Stone, Space of Contestation: Urban Conservation and the Colonial Past in Zanzibar." PhD diss., University of Chicago.

Black, George, ed. (1997) *Islam and Justice: Debating the Future of Human Rights in the Middle East and North Africa.* New York: Lawyers Committee for Human Rights.

Bowen, John R. (1988) "The Transformation of an Indonesian Property System: A*dat*, Islam and Social Change in the Gayo Highlands." *American Ethnologist* 15(2): 274–293.

——— (1996) " 'Religion in the Proper Sense of the Word': Law and Civil Society in Islamist Discourse." *Anthropology Today* 12(4): 12–14.

——— (1998a) "Law and Social Norms in the Comparative Study of Islam." *American Anthropologist* 100(4): 1034–1038.

——— (1998b) "Qur'an, Justice, Gender: Internal Debates in Indonesian Islamic Jurisprudence." *History of Religions* 38(1): 52–78.

——— (1998c) " 'You May Not Give It Away': How Social Norms Shape Islamic Law in Contemporary Indonesian Jurisprudence." *Islamic Law and Society* 5(3): 382–408.

——— (1999a) "Legal Reasoning and Public Discourse in Indonesian Islam." In Dale F. Eickelman and Jon W. Anderson, eds., *New Media in the Muslim World: The Emerging Public Sphere.* Bloomington: Indiana University Press: 80–105.

——— (1999b) "Modern Intentions: Reshaping Subjectivities in an Indonesian Muslim Society." In Robert W. Hefner and Patricia Horvatich, eds., *Islam in an Era of Nation States: Politics and Religious Renewal in Muslim Southeast Asia.* Honolulu, University of Hawaii Press: 157–181.

——— (2000) "Consensus and Suspicion: Judicial Reasoning and Social Change in an Indonesian Society 1960–1994." *Law and Society Review* 34(1): 97–128.

——— (2003) *Islam, Law and Equality in Indonesia: An Anthropology of Public Reasoning.* Cambridge: Cambridge University Press.

Brison, Karen J. (1999) "Imagining a Nation in Kwanga Village Courts, East Sepik Province, Papua New Guinea." *Anthropological Quarterly* 72(2): 74–85.

Burgess, Thomas (1999) "Remembering Youth: Generation in Revolutionary Zanzibar." *Africa Today* 46(2): 29–50.

Buskens, Léon (2008) "Tales According or the Book: Professional Witnesses ('*udul*) as Cultural Brokers in Morocco." In Baudouin Dupret, Barbara Drieskins, and Annelies Moors, eds., *Narratives of Truth in Islamic Law.* London and New York: I.B. Tauris: 143–160.

Caplan, Patricia (1975) *Choice and Constraint in Swahili Community.* London: Oxford University Press.

——— (1976) "Boys' Circumcision and Girls' Puberty Rites among the Swahili of Mafia Island, Tanzania." *Africa* 46(1): 21–33.

——— (1977) *African Voices, African Lives.* London, New York: Routledge.

——— (1984) "Cognatic Descent, Islamic Law and Women's Property on the East African Coast." In Renee Hirschon, ed., *Women and Property, Women as Property.* New York: St. Martin's Press: 23–43.

——— (1989) "Perceptions of Gender Stratification." *Africa* 59(2): 196–208.

——— (1995) " 'Law' and 'Custom': Marital Disputes on Northern Mafia Island, Tanzania." In Patricia Caplan, ed., *Understanding Disputes: The Politics of Argument.* Berg: Oxford Providence: 203–221.

Carrol, Lucy (1996) "Qur'an 2:229: 'A Charter Granted to the Wife? Judicial *Khul'* in Pakistan."
Islamic Law and Society 3(1): 91–126.

Chanock, Martin (1985) *Law, Custom and Social Order: The Colonial Experience in Malawi and Zambia.* Cambridge: Cambridge University Press.

Chase, Hank (1976) "The Zanzibar Treason Trial." *Review of African Political Economy* 6: 14–33.

Christelow, Alan, ed. (1994) *Thus Ruled Emir Abbas: Selected Cases from the Records of the Emir of Kano's Judicial Council.* Africa Historical Sources Series. East Lansing: Michigan State University Press.

Collier, Jane F. (1975) "Legal Processes." *Annual Review of Anthropology* 4: 121–144.

Comaroff, John L. and Jean Comaroff, eds. (1999) *Civil Society and the Political Imagination in Africa.* Chicago: University of Chicago Press.

Comaroff, John L. and Simon Roberts (1981) *Rules and Processes: The Cultural Logic of Dispute in an African Context.* Chicago: University of Chicago Press.

Conley, John M. and William M. O'Barr (1990) *Rules versus Relationships: The Ethnography of Legal Discourse.* Chicago: University of Chicago Press.

Cooper, Frederick (1980) *From Slaves to Squatters: Plantation Labor and Agriculture in Zanzibar and Coastal Kenya: 1890–1925.* New Haven: Yale University Press.

Coulson, Noel J. (1964) *A History of Islamic Law.* Edinburgh: Edinburgh University Press.

Dennerlein, Bettina (1999) "Changing Conceptions of Marriage in Algerian Personal Status Law." In Ravindra Khare, ed., *Perspectives on Islamic Law, Justice and Society.* Lanham, MD: Rowman & Littlefield: 123–141.

Dupret, Baudouin (2006) "The Practice of Judging: The Egyptian Judiciary at work in a Personal Status Case." In Khalid Muhammad Masud, Rudolph Peters, and David Powers, eds., *Dispensing Justice in Islam: Qadis and Their Judgments.* Leiden: E.J. Brill: 143–168.

——— (2007) "What Is Islamic Law? A Praxiological Answer and an Egyptian Case Study." *Theory, Culture and Society* 24: 79–100.

Dupret, Baudouin, Barbara Drieskens, and Annelies Moors, eds. (2008) *Narratives of Truth in Islamic Law.* London: I.B. Tauris.

Duranti, Alessandro (1987) "Intentions, Language, and Social Action in a Samoan Context." *Journal of Pragmatics* 12: 13–33.

Dwyer, Daisy (1977) "Bridging the Gap between Sexes in Moroccan Judicial Practice." In A. Schlegel, ed., *Sexual Stratification: A Cross-Cultural Perspective.* New York: Columbia University Press.

——— (1979) "Law Actual and Perceived." *Law and Society Review* 13(3): 740–756.

Dwyer, Daisy, ed. (1990) *Law and Islam in the Middle East.* New York: Begin and Garvey.

Eastman, Carol M. (1984) "Waungwana and Wanawake: Muslim Ethnicity and Sexual Segregation in Coastal Kenya." *Journal of Multilingual and Multicultural Development* 5: 91–112.

——— (1988) "Women, Slaves and Foreigners: African Cultural Influences and Group Processes in the Formation of Northern Swahili Coastal Society." *International Journal of African Historical Studies* 21 (1): 1–20.

Echtler, Magnus (2006) "Recent Changes in the New Year's Festival in Makunduchi, Zanzibar: A Reinterpretation." In Roman Loimeierand Rudiger Sesseman, eds., *The Global Worlds of the Swahili: Interfaces of Islam, Identity and Space in 19th and 20th-Century East Africa.* Berlin: Lit Verlag: 131–159.

Eickelman, Dale (1985) *Knowledge and Power in Morocco: The Education of a Twentieth-Century Notable.* Princeton: Princeton University Press.

Elmasri, F.H. (1987) "Sheikh al-Amin bin Ali al-Mazrui and the Islamic Intellectual Tradition in East Africa." *Journal Institute of Muslim Minority Affairs* 8(2): 229–237.

Ergene, Bogac (2004) "Pursuing Justice in an Islamic Context: Dispute Resolution in an Ottoman Court of Law." *Political and Legal Anthropology Review* 27(1): 67–87.

Esposito, John L. (1982) *Women in Muslim Family Law.* Syracuse: Syracuse University Press.

Ewing, Katherine, ed. (1988) *Shari'at and Ambiguity in South Asian Islam.* Berkeley: University of California.

Fair, Laura (2001) *Pastimes & Politics: Culture, Community and Identity in Post-Abolition Urban Zanzibar 1890–1945.* Athens: Ohio University Press and James Currey.

Fallers, Lloyd A. (1969) *Law without Precedent.* Chicago: University of Chicago Press.

Farsy, Abdallah Saleh (1965) *Ndoa, Talaka na Maamrisho Yake.* Zanzibar: Mulla Karimjee Mulla Mohamedbhai.

——— (1989) *The Shafi'i Ulama of East Africa 1830–1970: A Hagiographic Account.* Madison, African Studies Program: University of Wisconsin.

Flint, John E. (1965) *Zanzibar: 1890–1950.* In Low, D.A., E.M. Chilver, and Alison Smith, eds., *History of East Africa.* Oxford: Clarendon Press: 641–671.

Fluehr-Lobban, Caroline (1987) *Islamic Law and Society in the Sudan.* London: Frank Cass.

Fluet, Edward R., Mark J. Calaguas, and Cristina M. Drost (2006) "Legal Pluralism & Women's Rights: A Study in Post-Colonial Tanzania." Working Paper 1683, Berkeley Electronic Press.

Geertz, Clifford (1983) *Local Knowledge: Further Essays in Interpretive Anthropology.* New York: Basic Books.

Gerber, Haim (1999) *Islamic Law and Culture 1600–1840.* Leiden: E.J. Brill.

Glassman, Jonathan (1991) "The Bondsman's New Clothes: The Contradictory Consciousness of Slave Resistance on the Swahili Coast." *Journal of African History* 32: 277–312.

Gleave, Robert and Eugenia Kermeli, eds. (2001) *Islamic Law: Theory and Practice.* London and New York: I.B. Tauris.

Goody, Jack (1973) "Bridewealth and Dowry in Africa and Eurasia" in Jack Goody and Stanley J. Tambiah, eds., *Bridewealth and Dowry.* Cambridge: Cambridge University Press: 1–58.

Gower, Rebecca and Steven Salm (1996) "Swahili Women Since the Nineteenth Century: Theoretical and Empirical Considerations on Gender and Identity Construction." *Africa Today* 43(3): 251–269.

Gray, John Milner (1962) *History of Zanzibar from the Middle Ages to 1856.* London: Oxford University Press.

——— (1963) "Zanzibar and the Coastal Belt, 1840–1884." In R. Oliver and G. Mathew, eds., *History of East Africa,* vol. 1. Oxford: Clarendon Press: 212–251.

——— (1977) "The Hadimu and Tumbatu of Zanzibar." *Tanzanian Notes and Records* 81–82: 135–153.

Griffiths, Anne M.O. (1997) *In the Shadow of Marriage: Gender and Justice in an African Community.* Chicago: University of Chicago Press.

——— (2000) "Gender, Power and Difference: Reconfiguring Law from Bakwena Women's Perspectives." *PoLAR: Political and Legal Anthropology Review* 23(2): 89–106.

Grimes, Jill (1993) "Women, the Family and Marriage Laws in Zanzibar." UNICEF Report.

Habermas, Jurgen (1996) *Between Facts and Norms.* Cambridge, MA: MIT Press.

Haeri, Shahla (1989) "Ambivalence toward Women In Islamic Law and Ideology." *Middle East Annual* 5: 45–67.

——— (1989) *Law of Desire.* Syracuse: Syracuse University Press.

Hallaq, Wael (1997) *A History of Islamic Legal Theories.* Cambridge: Cambridge University Press.

——— (2001) "From Regional to Personal Schools of Law?" *Islamic Law and Society* 8(1): 1–26.

Hashim, Abdulkadir (2005) "Muslim Personal Law in Kenya and Tanzania: Tradition and Innovation." *Journal of Muslim Minority Affairs* 25(3): 449–460.

Haviland, John B. (1995) "Texts from Talk in Tzotzil." In Michael Silverstein and Greg Urban, eds., *Natural Histories of Discourse.* Chicago: University of Chicago Press: 45–79.

Hirsch, Susan F. (1994) "*Kadhi*'s Courts as Complex Sites of Resistance: The State, Islam and Gender in Postcolonial Kenya." In Mindie Lazarus-Black and Susan F. Hirsch, eds., *Contested States: Law, Hegemony and Resistance*. New York: Routledge: 207–230.

———— (1998) *Pronouncing and Persevering: Gender and the Discourse of Disputing in an African Islamic Court*. Chicago: University of Chicago Press.

———— (2006) "Islamic Law and Society Post 9/11." *Annual Review of Law and Social Science* 2: 165–186.

Hirschon, Renee (1984) *Women and Property, Women as Property*. New York: Palgrave Macmillan.

Hollander, Isaac (1995) "Ibra in Highland Yemen: Two Jewish Divorce Settlements." *Islamic Law and Society* 2(1): 1–9.

Hollingsworth, L.W. (1975) *Zanzibar under the Foreign Office 1890–1913*. Westport: Greenwood Press.

Ingrams, W.H. (1931, 1967) *Zanzibar, Its History and People*. London: Frank Cass.

———— (1962) "Islam and Africanism in Zanzibar." *New Commonwealth* 40(7): 427–430.

Jackson, Sherman (2001) "Kramer versus Kramer in an 18th/16th Century Egyptian Court: Post-formative Jurisprudence between Exigency and Law." *Islamic Law and Society* 8(1): 27–51.

Jones, Christina (1995) "Plus ca Change, Plus ca Reste la Meme? The New Zanzibar Land Law Project." *Journal of African Law* 40(1): 19–42.

Jones-Pauly, Christina and Stefanie Elbern (2002) "Access to Justice: The Role of Court Administrators and Lay Adjudicators in the African and Islamic Contexts." The Hague: Kluwer Law International.

Just, Peter (1990) "Dead Goats and Broken Betrothals: Liability and Equity in Dou Dongo Law." *American Ethnologist*: 75–90.

———— (2001) *Dou Dongoo Justice: Conflict and Morality in an Indonesian Society*. Lanham, MD: Rowman & Littlefield.

Kamali, Muhammad Hashim (1991) *Principles of Islamic Jurisprudence*. Cambridge: Islamic Texts Society.

Khare, Ravindra S., ed. (1999) *Perspectives on Islamic Law, Justice, and Society*. Lanham, MD: Rowman & Littlefield.

Knappert, Jan (2001) *Law Glossary of Islamic Terms in Swahili*. Tanzania: Benedictine.

Laitin, David (1982) "The Shari'ah Debate and the Origins of Nigeria's Second Republic." *Journal of Modern African Studies* 20(3): 411–430.

Lambek, Michael (1990) "Certain Knowledge, Contestable Authority: Power and Practice on the Islamic Periphery." *American Ethnologist* 17(1): 23–40.

Landberg, Pamela (1986) "Widows and Divorced Women in Swahili Society." In Betty Potash, ed., *Widows in African Societies: Choices and Constraints*. Stanford: Stanford University Press: 107–130.

Layish, Aharon (1975) *Women and Islamic Law in a Non-Muslim State*. Jerusalem: Israel Universities Press.

———— (1988) "Customary *Khul* as Reflected in the *Sijill* of the Libyan Shari'ah Courts." *Bulletin of the School of Oriental and African Studies* 51(3): 428–439.

———— (1991) *Divorce in the Libyan Family*. New York: New York University Press.

Lazarus-Black, Mindie and Susan F. Hirsch, eds. (1994) *Contested States: Law, Hegemony and Resistance*. New York: Routledge.

Le Guennec-Coppens, Francoise (1980) *Wedding Customs in Lamu*. Nairobi: Lamu Society.

Libson, Gideon (1997) "On the Development of Custom as a Source of Law in Islamic Law." *Islamic Law and Society* 4(2): 131–155.

Lienhardt, Peter (1959) "The Mosque College of Lamu and Its Social Background." *Tanzania Notes and Records* 53: 1–40.

Lofchie, Michael F. (1965) *Zanzibar: Background to Revolution*. Princeton: Princeton University Press.

Loimeier, Roman and Rudiger Sesseman, eds. (2006) *The Global Worlds of the Swahili: Interfaces of Islam, Identity and Space in 19th and 20th-Century East Africa*. Berlin: Lit Verlag.

Low, D.A. and Alison Smith, eds. (1976) *History of East Africa*. Oxford: Clarendon Press.

Mahmood, Saba (2004) *Politics of Piety: The Islamic Revival and the Feminist Subject*. Princeton: Princeton University Press.

Mallat, Chilbi and Jane Conners, eds. (1995) *Islamic Family Law*. London: Graham and Trotman.

Mamdani, Mahmood (1995) *Citizen and Subject*. Princeton: Princeton University Press.

—— (2004) *Good Muslim, Bad Muslim: America, the Cold War and the Roots of Terror*. New York: Pantheon.

Martin, B.G. (1969) "Muslim Politics and Resistance to Colonial Rule: Shaykh Uways B. Muhammad Al-Barawi and the Qadiriya Brotherhood in East Africa." *Journal of African History* 10(3): 471–486.

—— (1971) "Notes on Some Members of the Learned Classes of Zanzibar and East Africa in the Nineteenth Century." *African Historical Studies* 4(3): 525–545.

Martin, Esmond Bradley (1978) *Zanzibar: Tradition and Revolution*. London: Hamish Hamilton.

Masud, Muhammad Khalid, Brinkley Messick, and David Powers, eds. (1996) *Islamic Legal Interpretation: Muftis and Their Fatwas*. Cambridge, MA: Harvard University Press.

Masud, Muhammad Khalid, Rudolph Peters, and David Powers, eds. (2006) *Dispensing Justice in Islam: Qadis and their Judgments*. Leiden: E.J. Brill.

Mazrui, Alamin M. a. I.N. Shariff. (1994) *The Swahili: Idiom and Identity of an African People*. Trenton: Africa World Press.

Merry, Sally Engle (1988) "Legal Pluralism." *Law and Society Review* 20(5): 869–896.

—— (1990) *Getting Justice and Getting Even: Legal Consciousness among Working Class Americans*. Chicago: University of Chicago Press.

—— (1991) "Law and Colonialism." *Law and Society Review* 25(4): 889–922.

—— (1992) "Anthropology, Law, and Transnational Processes." *Annual Review of Anthropology* 21: 357–379.

Messick, Brinkley (1986) "The Mufti, the Text and the World: Legal Interpretation in Yemen." *JRAI: Man* 21(1): 102–119.

—— (1990) "Literacy and the Law: Documents and Document Specialists in Yemen." In Daisy Dwyer, ed., *Law and Islam in the Middle East*. New York: Bergin and Garvey: 61–76.

—— (1993) *The Calligraphic State: Textual Domination and History in a Muslim Society*. Berkeley: University of California Press.

—— (1996) "Media Muftis: Radio Fatwas in Yemen. Islamic Legal Interpretation: Muftis and Their Fatwas." In Muhammad Khalid Masoud, Brinkley Messick, and David Powers, eds., *Islamic Legal Interpretation: Muftis and Their Fatwas*. Cambridge, MA: Harvard University Press: 310–322.

—— (1998) "Written Identities: Legal Subjects in an Islamic State." *History of Religions* 38(1): 25–51.

—— (2001) "Indexing the Self: Intent and Expression in Islamic Legal Acts." *Islamic Law and Society*, 8: 151–178.

—— (2002) "Evidence: From Memory to Archive." *Islamic Law and Society* 9: 231–270.

—— (2008) "Legal Narratives from Shari'a Courts." In Baudouin Dupret, Barbara Drieskens, and Annelies Moors, eds., *Narratives of Truth in Islamic Law*. London: I.B. Tauris: 51–68.

Middleton, John (1961) *Land Tenure in Zanzibar*. London: His Majesty's Stationary Office.

—— (1992) *The World of the Swahili: An African Mercantile Civilization*. New Haven: Yale University.

Middleton, John and Jane Campbell (1965) *Zanzibar: Its Society and Politics*. Oxford: Oxford University Press.

Mir-Hosseini, Ziba (1992/1993) "Paternity, Patriarchy and Matrifocality in the Shari'a and in Social Practice: The Cases of Morocco and Iran." *Cambridge Anthropology* 16(2): 22–40.

——— (1993) "Women, Marriage and the Law in Post-Revolutionary Iran." In Haleh Afshar, ed., *Women in the Middle East: Perceptions, Realities and Struggles for Liberation*. New York: St. Martin's Press: 59–84.

——— (1999) *Islam and Gender: The Religious Debate in Contemporary Iran*. Princeton: Princeton University Press.

——— (2000) *Marriage on Trial: A Study of Islamic Family Law; Iran and Morocco Compared*. London and New York: I.B. Tauris.

Mirza, Sarah and Margaret Strobel (1989) *Three Swahili Women: Life Histories from Mombasa, Kenya*. Bloomington: Indiana University Press.

Moore, Erin P. (1993) "Gender, Power and Legal Pluralism: Rajasthan, India." *American Ethnologist* 20(3): 522–542.

——— (1998) *Gender, Law and Resistance in India*. Tucson: University of Arizona Press.

Moore, Sally Falk (1978) *Law as Process*. London: Routledge and Kegan Paul.

——— (1986) *Social Facts and Fabrications: "Customary Law" on Kilimanjaro, 1880–1980*. Cambridge: Cambridge University Press.

——— (1989) "History and the Redefinition of Custom on Kilimanjaro." In June Starr and Jane F. Collier, eds., *History and Power in the Study of Law*. Ithaca and London: Cornell University Press.

Moors, Annelies (1995) *Women, Property and Islam: Palestinian Experiences: 1920–1990*. Cambridge: Cambridge University Press.

——— (1999) "Debating Islamic Family Law: Legal Texts and Social Practices." In Margaret L. Meriwether and Judith E. Tucker, eds., *Social History of Women and Gender in the Modern Middle East*. Boulder: Westview Press: 141–176.

——— (2008) "Registering a Token Dower: The Multiple Meanings of a Legal Practice." In Dupret, Baudouin, Barbara Drieskens, and Annelies Moors, eds., *Narratives of Truth in Islamic Law*. London: I.B. Tauris: 85–104.

Mundy, Martha (1995) *Domestic Government: Kinship, Community and Polity in North Yemen*. London: I.B. Tauris.

Nader, Laura and Barbara Yngvesson (1973) "On Studying the Ethnography of Law and Its Consequences," *Handbook of Social and Cultural Anthropology*, 883–921.

Nuotio, Hanni (2006) "The Dance That Is Not Danced, the Song That Is Not Sung: Zanzibari Women in the *Maulidi* Ritual." In Roman Loimeier and Rudiger Sesseman, eds., *The Global Worlds of the Swahili: Interfaces of Islam, Identity and Space in 19th and 20th-Century East Africa*. Berlin: Lit Verlag: 187–208.

Parkin, David (2000) "Islam among the Humors: Destiny and Agency among the Swahili." In Ivan Karp and D.A. Masolo, eds., *African Philosophy as Cultural Inquiry*. Bloomington: Indiana University Press: 50–65.

Pearce, Frances Barrow (1920, 1967) *Zanzibar: The Island Metropolis of Eastern Africa*. London: Frank Cass.

Peirce, Leslie (1998) " 'She Is Trouble . . . and I Will Divorce Her': Orality, Honor and Representation in the 16th-Century Ottoman Court of 'Aintab." In Gavin R.G. Hambly, ed., *Women in the Medieval Islamic World: Power, Patronage and Piety*. New York: St. Martin's Press: 269–300.

——— (2003) *Morality Tales: Law and Gender in the Ottoman Court of Aintab*. Berkeley: University of California Press.

Peletz, Michael (2002) *Islamic Modern: Religious Courts and Cultural Politics in Malaysia*. Princeton: Princeton University Press.

Pouwels, Randall L. (1981) "Sh. al-Amin b. Ali Mazrui and Islamic Modernism in East Africa 1875–1947." *International Journal of Middle East Studies* 13: 329–345.

——— (1987) *Horn and Crescent: Cultural Change and Traditional Islam on the East African Coast 800–1900.* Cambridge: Cambridge University Press.

Purpura, Allyson (1997) "Knowledge and Agency: The Social Relations of Islamic Expertise in Zanzibar Town." PhD diss., City University of New York.

Ramadhani, A.S.L. (1978–1981) "Judicial System of Tanzania, Zanzibar." *East African Legal Review* 11–14: 225–239.

Rapoport, Yosef (2000) "Matrimonial Gifts in Early Islamic Egypt." *Islamic Law and Society* 7(1): 1–36.

Reese, Scott S., ed. (2004) *The Transmission of Learning in Islamic Africa.* Leiden: E.J. Brill.

Reiter, Yitzhak (2001) "Qadis and the Implementation of Islamic Law in Modern Day Israel." In Robert Gleave and Eugenia Kermeli, eds., *Islamic Law: Theory and Practice.* London and New York: I.B. Tauris: 205–231.

Roberts, Simon (1994) "Law and Dispute Processes." In Tim Ingold, ed., *Companion Encyclopedia of Anthropology: Humanity, Culture and Social Life.* London and New York: Routledge: 962–982.

Rosaldo, Michelle (1982) "The Things We Do with Words: Ilongot Speech Acts and Speech Act Theory in Philosophy." *Language and Society* 2: 203–237.

Rosen, Lawrence (1984) *Bargaining for Reality: The Construction of Social Relations in a Muslim Community.* Chicago: University of Chicago Press.

——— (1989) *The Anthropology of Justice.* Cambridge: Cambridge University Press.

——— (1995a) "Law and Custom in the Popular Legal Culture of North Africa." *Islamic Law and Society* 2(2): 194–208.

———, ed. (1995b) *Other Intentions.* Santa Fe: School of American Research.

——— (1999) "Justice in Islamic Culture and Law." In Ravindra Khare, ed., *Perspectives on Islamic Law, Justice and Society.* Lanham, MD: Rowman & Littlefield: 34–52.

——— (2000) *The Justice of Islam.* Oxford: Oxford University Press.

Rwezaura, Barthazar A. and Ulrike Wanitzek (1988) "Family Law Reform in Tanzania: A Socio-Legal Report." *International Journal of Law and the Family* 2: 1–26.

Schacht, Joseph (1964) *An Introduction to Islamic Law.* Oxford: Clarendon Press.

Searle, John (1983) *Intentionality.* Cambridge: Cambridge University Press.

Seng, Yvonne J. (1994) "Standing at the Gates of Justice: Women in the Law Courts of Early Sixteenth-Century Uskudur, Istanbul." In Lazarus-Black, M. and S. Hirsch, eds., *Contested States: Law, Hegemony, and Resistance.* New York: Routledge: 184–206.

Shaham, Ron (1995) "Custom, Islamic Law, and Statutory Legislation: Marriage Registration and Minimum Age at Marriage in the Egyptian Shari'a Courts." *Islamic Law and Society* 2(3).

Shariff, O.M.O (forthcoming) *Zanzibar Legal System. Contemporary Zanzibar.* Nairobi: INFRA.

Sheriff, Abdul (1991) "The Peasantry under Imperialism, 1873–1963." In Abdul Sheriff and Ed Ferguson, eds., *Zanzibar under Colonial Rule.* London: James Currey and Ohio Univeristy Press: 109–140.

Sheriff, Abdul and Ed. Ferguson, eds. (1991) *Zanzibar Under Colonial Rule.* London: James Currey and Ohio University Press.

Silverstein, Michael and Greg Urban (1996) *Natural Histories of Discourse.* Chicago: University of Chicago Press.

Simmons, A. John (1999) "Fault, Objectivity, and Classical Islamic Justice." In Ravindra Khare, ed., *Perspectives on Islamic Law, Justice and Society.* Lanham, MD: Rowman & Littlefield: 52–61.

Smith, Alison (1976) "The End of the Arab Sultanate: Zanzibar 1945–1964." In Low, D.A. and
    Alison Smith, eds., *History of East Africa*, vol. 3. Oxford: Clarendon Press: 199.
Sonbol, Amira el Azhary, ed. (1996) *Women, the Family, and Divorce Laws in Islamic History.*
    Syracuse: Syracuse University Press.
———— (2005) *Beyond the Exotic: Women's Histories in Islamic Societies.* Syracuse: Syracuse
    University Press.
Starr, June (1978) *Dispute and Settlement in Rural Turkey.* Leiden: E.J. Brill.
———— (1990) "Islam and the Struggle over State Law in Turkey." In Daisy Dwyer, ed., *Law and
    Islam in the Middle East.* New York: Bergin and Garvey: 77–98.
———— (1992) *Law as Metaphor: From Islamic Courts to the Palace of Justice.* Binghamton: SUNY
    Press.
———— (1993) "When Empires Meet: European Trade and Ottoman Law." In Mindie Lazarus-
    Black and Susan Hirsch, eds., *Contested States: Law, Hegemony and Resistance.* New York:
    Routledge: 231–251.
Stiles, Erin Elizabeth (2003) "When Is a Divorce a Divorce? Determining Intention in Zanzibar's
    Islamic Courts." *Ethnology* 42(4): 273–288.
———— (2005) " 'There Is No Stranger to Marriage Here!' Muslim Women and Divorce in
    Rural Zanzibar." *Africa* 75(4): 582–598.
———— (2006) "Broken Edda and Marital Mistakes: Two Recent Disputes from an Islamic
    Court in Zanzibar." In Muhammad Khalid Masud, Rudolph Peters, and David Powers, eds.,
    Dispensing Justice in Islam: Qadis and Their Judgments. Leiden: E.J. Brill: 95–116.
Stockreiter, Elke Elisabeth (2008) "Tying and Untying the Knot: Kadhi's Courts and the Negotiation
    of Social Status in Zanzibar Town: 1900–1963." PhD diss., University of London.
Stoeltje, Beverly (2002) "Performing Litigation at the Queen Mother's Court." In Christina
    Jones-Paulyand Stefanie Elbern, eds., *Access to Justice: The Role of Court Administrators and Lay
    Adjudicators in the African and Islamic Contexts.* Leiden: E.J. Brill: 1–22.
Strobel, Margaret (1979) *Muslim Women in Mombasa, 1890–1975.* New Haven: Yale University
    Press.
Swartz, Marc J. (1979) "Religious Courts, Community and Ethnicity among the Swahili of
    Mombasa: An Historical Study of Social Boundaries." *Africa* 49(1): 29–40.
Trimingham, John Spencer (1944) *Islam in East Africa.* Oxford: Clarendon Press.
Tucker, Judith E. (1998) *In the House of the Law: Gender and Islamic Law in Ottoman Syria and
    Palestine.* Berkeley: University of California Press.
Vaughn, J.H. (1935) *The Dual Jurisdiction in Zanzibar.* Zanzibar: Government Printer.
Vogel, Frank (1993) "Islamic Law and Legal System Studies of Saudi Arabia." PhD diss., Harvard
    Univeristy.
Watson, C.W. (1992/1993) "Islamic Family Law and the Minangkabau of West Sumatra."
    Cambridge Anthropology 16(2): 69–83.
Welchman, Lynn (2000) *Beyond the Code: Muslim Family Law and the Shari'a Judiciary in the
    Palestinian West Bank.* Boston: Kluwer Law International.
———— (2004) *Women's Rights and Islamic Family Law: Perspectives on Reform.* New York: Zed
    Books.
Wensinck, A.J. (1987) "*Niyya.*" In *Encyclopedia of Islam.* Leiden: E.J. Brill.
Wurth, Anna (1995) "A Sana'a Court: The Family and the Ability to Negotiate." *Islamic Law
    and Society* 2(3): 320–340.
Yngvesson, Barbara (1993) *Virtuous Citizens, Disruptive Subjects: Order and Complaint in a New
    England Court.* New York: Routledge.

INDEX